Fishing for Gold

Fishing for Gold

The Story of Alabama's Catfish Industry

Karni R. Perez

THE UNIVERSITY OF ALABAMA PRESS
Tuscaloosa

Manufactured in the United States of America

Typeface: Minion

∞

The paper on which this book is printed meets the minimum requirements of American National Standard for Information Science—Permanence of Paper for Printed Library Materials, ANSI Z39.48-1984.

Library of Congress Cataloging-in-Publication Data

Perez, Karni R.
Fishing for gold : the story of Alabama's catfish industry / Karni R. Perez.
p. cm.
"Fire Ant book"
Includes bibliographical references and index.
ISBN-13: 978-0-8173-5344-5 (pbk. : alk. paper)
ISBN-10: 0-8173-5344-5
1. Catfishes—Alabama—History. 2. Fish culture—Alabama—History. I. Title.
SH167.C35P47 2006
639.3′7492′09761—dc22

2005015400

Contents

Illustrations

Map

Figures

Acknowledgments

I am grateful to all the individuals who welcomed me into their homes and offices; gave me their time, attention, and assistance; and shared with me recollections and information from their early involvement in Alabama's catfish industry. The insight they provided me on their experiences and their approaches to their lives made an indelible impression and brought to life for me the events described here. I hope I have done justice to their character and accomplishments. I am indebted to the reviewers for the generous time and effort they spent pointing out errors and weaknesses, and ways in which the manuscript could be improved. I owe thanks also to the many individuals who provided facts that were necessary to help fill out the total picture presented. Permission to use sections of copyrighted publications for epigraphs was granted by the Catfish Institute; the U.S. Trout Farmers Association; Gary C. Cox; Porter Briggs of *Commercial Fish Farmer;* and Robert Fisher of *American Fish Farmer.*

I owe personal thanks to many for always being there for me with their support and faith in my ability to do a good job. On innumerable occasions, their help and understanding were central to my working through the knots that formed throughout the project. Among them are, especially, Conner Bailey, Check Stephens, and Jay Lamar. I could not have done any of this without the never-ending patience and encouragement of my family—my husband, Joe, my sons, Mike and Josh, and my dog, Roger.

Work on this project was supported jointly by the statewide farmers' organization, the Alabama Catfish Producers, a division of the Alabama Farmers Federation; the leading Alabama processor, Southern Pride; and by the Department of Fisheries and Allied Aquacultures at Auburn University.

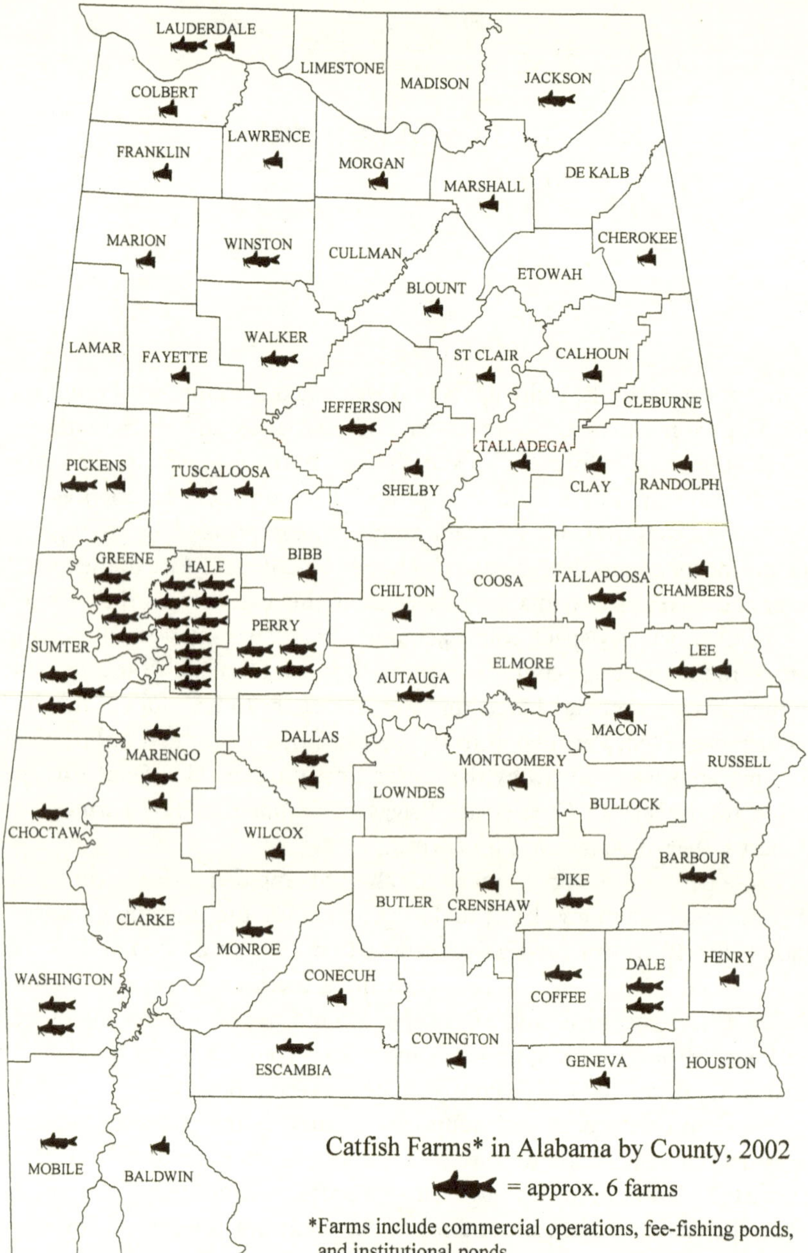

Catfish Farms* in Alabama by County, 2002

[catfish symbol] = approx. 6 farms

*Farms include commercial operations, fee-fishing ponds, and institutional ponds.

Source: U.S. Department of Agriculture, National Agricultural Statistics Service, 2002 Census of Agriculture

Introduction

In the history of human endeavors, fish culture is not new. In parts of Asia, people have been raising fish for food for thousands of years, and European experience goes back centuries. The channel catfish, however, is a relative newcomer to fish culture even in the United States. As a commercial pursuit, culture of catfish did not get a firm foothold until around 1960. But when it did, it developed into an enterprise at which thousands of farmers and entrepreneurs tried their hands. In contrast to some attempts to establish fish-growing enterprises earlier in the twentieth century, catfish farming and processing succeeded beyond many people's wildest dreams.

Fish farmers in Arkansas, with experience in managing water for rice cultivation and growing baitfish, began raising catfish commercially before Alabama farmers entered the industry. But most people think of Mississippi when the words "catfish farming" or "catfish industry" are used today. Shortly after Alabamians began raising catfish commercially, a fledgling industry arose in Mississippi and soon overtook all other states in production. The Mississippi Delta region was a natural fit for fish farming. Flat land and abundant water-retaining "gumbo" soils of the Delta made it possible to locate ponds almost anywhere. A shallow water table allowed growers to rely on plentiful, accessible groundwater.

Historical and socioeconomic factors also contributed to the industry's rapid rise in that region. The Delta had a history of large-scale farming on huge acreage owned by relatively few people. The approach of Mississippi's landowners to farming in general may well be related to that history. From the beginning, catfish farming was practiced there on a large scale. Most fish farmers had at least a thousand acres of land, and some as many as ten thousand. They tended to think in big terms and be open to taking economic risk.[1] Research has confirmed that wealthier farmers with larger operations tend to be more progressive and more likely to adopt new methods than those with smaller ones, who may feel less protected from risks.[2]

With their real wealth and relatively large operations, Delta farmers had ready access to the sizable investment capital needed for the changeover to catfish farming on a large scale. They already owned equipment for farming and pond building and already had access to management and labor resources. Starting a fish farm was therefore less costly, and they were able to operate more efficiently from the beginning.[3] Also, political power in Mississippi resided historically in the Delta, and owners of large, successful fish farms were able to bring influence to bear on systems of government. State officials provided major support to the farmers there early on and continued to do so for years.[4]

According to John Jensen, of the Department of Fisheries and Allied Aquacultures at Auburn University, a key ingredient to the catfish industry's impressive growth in general has been a combination of natural conditions, public and private infrastructure, and people. All of these, together, have enabled and supported it. Among them, suitable soils and terrain, adequate water resources, and a climate that is warm for much of the year are basic. The channel catfish, itself, has essential qualities that not only make it suitable for culture but also provide an advantage over other species. Primarily, it has a large stomach and can hold feed for hours, so it is able to thrive taking feed just once a day. Additionally, though originally a river fish, it has shown it is capable of learning new behaviors that allow it to survive in commercially managed ponds. The fish used to be thought of by many as "dumb," but that has changed: "He's one of the smartest fish out there now. Naturally, he's a bottom fish. You train him to feed on top. If oxygen goes out, he'll go to that paddle wheel. A bass or bream will lay right over there on that bank and die. But a doggone catfish, you can train him just like any other kind of animal you can train."[5]

Services provided by public infrastructure such as good transportation systems, a supportive legal and regulatory foundation, research, outreach, and education are important parts of the mix. Private businesses providing financing, equipment, supplies, energy, and many other materials and services, as well as tradesmen and professionals, are all indispensable to modern farms and processing plants. Last, but not least, are the producers (including farmers and processors) and the consumers, both as a group and as individuals.

Alabama's portion of these key ingredients is not equivalent, in some regards, to that of Mississippi. Yet it has other assets from which it has benefited, such as its extensive river system; its use of rainfall runoff, "free" water, rather than groundwater; lower power costs; and Auburn University's long history of fisheries research and its scientists' working to provide the state's farmers with real solutions to their problems. Perhaps one of the most special assets of the Alabama industry is its people. Alabama's farmers and processors have shown themselves to be capable of amazing success as a group. Alabama has also benefited from the efforts of individuals—individuals whose vision, conviction, energy, hard work, patience, and ingenuity have driven the industry's progress.

Why look at Alabama for a history of the catfish industry? The industry in Alabama has developed differently from that in Mississippi. The path it took is noteworthy and its contributions are significant. Its story can teach some lessons about the process of developing agriculture. Few development efforts have been as successful as this grassroots one has been. This history illustrates some important ingredients for success in building a modern-day agricultural industry from the ground up.

The birthplace, so to speak, of the Alabama catfish industry was in the state's Black Belt. Alabama's Black Belt, named for its dark, fertile soils, lies in the western part of the state and is well known for the chronic poverty of many of its residents. It is a rural region once covered with cotton plantations, until the boll weevil made widespread cotton cultivation economically unfeasible early in the twentieth century. Black Belt farmers tried a number of other enterprises over the years since—among them row crops, soybeans, peanuts, and dairy and beef cattle. Some were successful for a time, but none brought with it a solution to the area's stubbornly dragging economy. Nothing on the horizon offered hope for significant improvement in jobs and income. In 1960, almost by accident, a small seed took root in Greensboro, a relatively quiet little town of about twenty-five hundred on the Black Belt's northern edge. It was the seed of an industry whose impacts would eventually grow and spill over into commercial fish farming around the world.

Few people realize that the success of the entire industry is based in significant measure to groundwork laid in Alabama, and credit is due. One of the purposes of this book is to illustrate the contributions of Alabamians who played a role in that groundwork. It is primarily an oral history, a retelling of the stories of farmers, processors, researchers, extensionists, and others. I have tried to let the people speak for themselves, because they can tell their stories best.

This book is an attempt to make the history readable and reachable to a wider audience than just those who already know a lot about the industry. For me, history is most immediate when the main focus is on the personal side. Perhaps these stories will help the general public gain a better understanding and appreciation of where their catfish meal came from and how it got to their table, historically speaking. Most of us today are so far removed from agriculture and food production that we know little about what has taken place in order for us to have the basic necessities of life, much less about the people responsible.

Many of the events described in the book took place almost two generations ago, and some of the individuals who were the actors at that time are no longer with us. If the story is not told now, much of it will disappear, never to be retrieved. Most of the people I interviewed, especially the farmers and others whose roles were not tied to an institution such as a government agency or a university, kept little in the way of records of their experiences. Most admitted to having had no idea their activities would ever amount to something anyone would want to know or write about later. They were too busy just figuring out what they needed to do next and then doing it.

One facet of the industry that could not be fully explored here is the experience of farm and plant workers. Plant workers, especially, are becoming an increasingly significant determinant in the future of the industry. Their stories will be a valuable contribution to understanding not only a sector of our society that has often been ignored in written history but also what makes the industry tick.

Researching and writing this history has taught me many things. I learned that history is a very difficult thing to write well. I learned that it is a great responsibility to report events in such a way that fact and truth are perpetuated. I learned that any single history can tell only part of the story. And I learned that there is no telling what individuals and groups can accomplish given a positive spirit, a strong community, and the necessary resources.

Notes

1. Richard True, interview by author, tape recording, 26 February 1997; Claude E. Boyd, interview by author, tape recording, 14 April 1998; Elvin Wright, interview by author, tape recording, 12 May 1998, all in Special Collections and Archives, Auburn University, Auburn, Ala.

2. Gary C. Cox, "The Catfish Farming Industry of Southeastern United States:

A Geographic Inquiry into the Patterns and Processes of Regional Aquaculture" (Ph.D. diss., University of Northern Colorado, 1972), 42–43.

3. Ramon Callahan, "There's Gold in the Catfish Country," *American Fishes and U.S. Trout News* 13 (September–October 1968): 20.

4. True, interview.

5. A. L. (Buddy) Lavender, interview by author, tape recording, 29 March 1998, Special Collections and Archives, Auburn University, Auburn, Ala.

1
Foundations

> Aquaculture is defined as the controlled cultivation and harvest of aquatic animals and plants.
>
> —Alabama Aquaculture Plan

The catfish industry as we know it did not arise out of a vacuum. Practices thousands of years ago in distant parts of the world laid the earliest groundwork for its comparatively recent, sudden rise as a significant economic player in Alabama, nationally, and globally. Through millennia, ideas of fish culture evolved with human experience, knowledge, and experimentation to make possible the industry as it exists today. The earliest fish known to be cultured by humans were carp. Written records show that these warmwater fishes were cultivated in China as long ago as the eleventh century B.C. and may have been raised for food as early as 2000 B.C. Out of the Chinese experience came the first known written description of fish culture around 475 B.C. Other Asian countries also have fairly long histories of fish culture. Evidence suggests that tilapia were farmed in ancient Egypt and that ancient Romans raised mullet for food. European experience goes back centuries to when the practice spread to Europe during the Middle Ages. Records indicate fish culture existed in the fifteenth century in Great Britain, where culturing methods were apparently advanced, and during the eighteenth century in Germany, where production of trout, a coldwater sport fish, had begun.

In North America, fish cultivation did not begin until the nineteenth century. Generally, the abundance of wild fish in native waters made it unnecessary to raise fish for food before that. When fish culture did begin in the United States, it was done primarily in order to enhance commercial fishing in public waters. Carp and trout culture started by the early to mid-1800s. Only later in that century were warmwater fish, including some catfish, first raised and distributed for stocking by federal agencies.

During the late 1800s the industrial economy and developments in transportation made Americans more mobile and increased their leisure time. As a result, emphasis began to shift from replenishing streams and lakes for commercial purposes to stocking them for recreational fishing. Federal and state fish hatcheries grew in number, and fish culture was viewed increasingly as worthy of scientific investigation. Federal research projects sprang up and led to observations on channel catfish culture and reproductive behavior in Kansas and Washington, D.C. Generally speaking, however, catfish were still being raised from captured wild eggs or immature wild stock rather than cultivated.

After 1900 channel catfish and other species received increasing attention from federal fish culturists, and gradually, more was learned about their reproduction (spawning) and rearing. Work in Kansas led to publications in 1881 and later that promoted the notion of pond construction for fish culture, but many early efforts to produce and stock fish were unsuccessful because of a lack of understanding about pond and stream ecology. The earliest known attempt to correct this came in 1914 when the Kansas Department of Game and Fish published one of the first manuals on pond fish culture in the United States. The manual contained guidelines for constructing ponds and culturing warmwater fish. In 1917 attempts to get channel catfish to spawn in a pond finally succeeded at the biological station of the U.S. Bureau of Fisheries in Iowa, where scientists had provided nesting sites in a pond, stocked brood fish in it, and found juveniles the following year.

While research was also carried out in Oklahoma and other states early in the century, Kansas continued to be one of the most active places for work with channel catfish. By 1921 fisheries experts at the State Fish Hatchery in Pratt advanced spawning methods and claimed they had achieved propagation of the fish. Between 1926 and 1929 they also made advances in hatching technology with the development of hatching troughs and paddle wheels to aerate the water in the troughs, helping pave the way for commercial production of fingerlings. Troughs and paddle wheels, their original design almost unchanged, are still used in the industry today.

By the late 1920s only a handful of farms in the United States were raising warmwater fish for commercial purposes. It was at about that time that the first warmwater fish farming industry began taking shape in the Southeast. Noting the increasing interest in sportfishing and encouraged by government wildlife and conservation agencies, farmers in Arkansas began raising minnows for bait and building their experience in pond fish production.

In 1931, while commercial bait minnow production was growing, a group of professors at Auburn University in Alabama undertook to build a small lake for fishing and as a source of recreation for themselves and their families.[1] They constructed the pond and stocked it with fish. When the dam collapsed within less than a year, M. J. Funchess, a member of the group and dean of the College of Agriculture, directed Homer Swingle, a young entomologist in the Department of Zoology and Entomology, to find out what had gone wrong.

Swingle put together a written research proposal and in 1934 obtained funding through the Purnell Act of 1925, which provided for money to support research that would develop and improve life for the nation's rural populations. The project aimed to investigate the production of freshwater fishes for food. The Purnell Project, later called the Farm Ponds Project, was born. The first general objectives were to find ways to (1) manage water resources productively, (2) improve sportfishing, and (3) produce fish as a source of food. Ponds were built on agricultural experiment station land near the Auburn campus, and research began (see figs. 1 and 2). No catfish were involved at the outset. Most of the fish stocked initially in the ponds were bluegill and largemouth bass, species used primarily for sportfishing in lakes.

The 1930s brought other developments essential to freshwater aquaculture in the United States and the culture of channel catfish in particular. In 1933 the Public Works Administration began building a federal fish hatchery in Marion, Alabama. Two years later, the U.S. Department of Agriculture's Soil Conservation Service (SCS) was established. Through its soil scientists, engineers, and planning technicians, the SCS was able to help farmers and landowners choose sites as well as design and lay out ponds. Over the years, SCS assistance and support was to figure prominently into the building of most of the ponds in the United States.

The Farm Ponds Project expanded, meanwhile, with additional sources of funding. Swingle was able to hire undergraduates and to increase the types of ponds used for experiments. Increasing numbers of fish species were included in the growing research program, but only a few catfish were among them. In the process of building, stocking, and maintaining the new ponds, the researchers gained fundamental knowledge about dam engineering techniques, pond-raised fish populations, fish production, health and parasite control, pond ecology, pond management methods, weed-control techniques, and fish biology. Their findings were channeled over the follow-

Fig. 1. Constructing a dam and early ponds at Auburn, 1930s.
Building a dam between early Auburn research ponds with slip-scrapes, a labor-intensive method used before mechanized earthmovers were developed. To excavate, laborers lifted the back handles of the slip-scrapes to scoop up soil. They held the handles down to retain the soil while mule teams pulled the slip-scrapes to the dam site. There, workers lifted the handles to let the load slide off. (Used by permission of Auburn University Libraries, Ralph Brown Draughon Library, Special Collections and Archives.)

ing decades into publications and courses shared with professionals and the public. In 1938 information in one of the earliest reports by Swingle and E. V. Smith summarized what had been learned about stocking, fertilization, and weed control and would later prove to be fundamental to the development of commercial catfish farming.

Sparse financial resources during World War II limited fisheries research everywhere. In spite of limited funding, researchers at Auburn continued to study channel catfish and other species for sportfishing purposes. Information and experience piled up steadily. During the 1940s work by fish culturists in Nebraska, Missouri, and other state research sites added to the mo-

Fig. 2. Laborers excavating with slip-scrape and one mule, 1930s.
A closer view of a slip-scrape, mule, and laborers. (Used by permission of Auburn University Libraries, Ralph Brown Draughon Library, Special Collections and Archives.)

mentum with findings on spawning and rearing of channel catfish and identification and control of diseases.

More reservoirs and private ponds were built throughout the country with the support of federal money and SCS consulting services. Ponds were encouraged as tools for conserving soil and water, and rural families were stocking them with sport fish for their own use. Watching these trends, Homer Swingle recognized there would be a growing need for biologists and fish culturists trained in pond fisheries. By the late 1940s Auburn staff began teaching fishpond management principles to SCS biologists, who then took what they had learned to pond owners in their respective states. The faculty also initiated a fisheries curriculum in order to educate and train students who were later to become experts in their fields.

After World War II, groundbreaking advances were achieved in the stimulation and control of spawning of catfish. Kermit Sneed and Howard Clemens, researchers with the U.S. Fish and Wildlife Service (FWS) in Oklahoma, had induced breeding by injecting pituitary hormones into the fish and had succeeded in getting pairs to spawn while contained in individual aquaria. Experience with channel catfish in Nebraska, Wisconsin, and later in Texas confirmed the effectiveness of techniques of pond spawning with containers in pens in ponds. The notion of selective breeding in catfish for disease resistance had been introduced by that time, and the use of spawning pens and aquaria allowed control over the selection of breeding pairs rather than the random mate selection that occurred in open ponds.[2] After these results were achieved, the direction of warmwater aquaculture shifted toward production of catfish, albeit still mainly for uses other than food. Although the methodology for spawning and rearing catfish was becoming fairly well established, only a few small commercial ventures existed.[3]

After Auburn researchers found that catfish coexisted well with bass and bream (sunfish), many pond owners and reservoir managers in Alabama and the Southeast began adding catfish to the other fishes they were raising. Interest in outdoor recreational activities such as fishing continued to grow with increasing numbers of ponds and reservoirs. Raising fish specifically as a source of food remained a secondary benefit. Conditions in the 1940s were not yet ripe for the notion of warmwater fishes as a staple in the nation's diet to catch hold.

U.S. aquaculture took a big forward leap in the early 1950s. Observers have suggested that the existence of surplus cropland, submarginal land, and many available reservoirs; the need for a better crop rotation system in rice culture; and the high economic value of freshwater fish all contributed to an increase in fish farming. By then, nearly 2 million ponds are estimated to have been built in the United States.[4] Many of these ponds had originally been intended primarily for watering livestock. A large proportion were in Arkansas, which was especially suited to pond building because of its ample groundwater, water-retaining soil, and farmers' previous involvement in rice production. Demand for baitfish increased during the early 1950s, and bait minnow farming begun there twenty years earlier became the first major warmwater aquaculture practiced in the United States.

Pond management techniques were being passed on to increasing numbers of students going through the fisheries curriculum at Auburn—students who would later establish their own farming enterprises, work on fish farms, or contribute to fisheries research. Jack Snow, the first to receive a master's degree in the program, graduated in 1948 and began work at the federal hatchery in Marion in 1950. There, with Swingle's cooperation, he built a training program for FWS fish culturists and eventually became the facility's director. The Marion hatchery began producing catfish fingerlings for pond owners for the first time in 1954. Able now to get fingerlings from Marion, the Auburn researchers could more easily use catfish for experiments.

In the early 1950s Homer Swingle began considering the possibility and potential of producing food fish on a commercial scale. At the time, a successful trout industry in the northwestern United States was the major source of commercially produced food fish. But trout is a coldwater fish, and Swingle wanted to find a warmwater fish—one that not only took prepared feed and grew rapidly but could also survive relatively poor water quality. The first of Auburn's commercial efforts were with buffalo fish and spotted bullheads, one of the smaller catfish species. Results suggested, however, that neither was suitable for farming for a variety of reasons.

Ideas about fish farming traveled largely by word of mouth, from farmer to farmer. Federal agencies helped spread the word by participating in state efforts to promote fish farming. At a meeting in Arkansas in 1956, Ben Nelson of Anderson's minnow farm in that state presented a paper on his experience, mentored by Kermit Sneed, with spawning channel catfish in aquaria. Papers such as his helped pass information to an even wider audience. But broad-scale, commercial catfish production was not yet taking place. Alabama farmers still looked to their ponds as a source of recreation and food for their families.

~

In the middle of the Sputnik era of the 1950s, funding for fisheries research was relatively easy to obtain. Everyone was eager to invest money in research that would broaden the boundaries of science and technical know-how and strengthen the free world against its cold war enemies. A second large wave of warmwater aquacultural activity began during the 1950s when many Arkansas rice farmers began investing heavily in raising buffalo fish in their rice fields. Then, within less than a decade, the buffalo fish industry collapsed almost totally, as farmers found that there was little demand for the fish. During the late 1950s and early 1960s, successes in agricultural production generally created a problematic surplus that drove prices down. In some sectors, including baitfish farming in the South, oversupply caused economic collapse or near collapse, causing many minnow farmers to go out of business.

Under these conditions, support for fish production research suddenly became difficult to find. Homer Swingle was convinced, however, that holding back research on a possible food source was too narrow a viewpoint, and he continued to look for ways to use aquaculture to provide better, more easily available nutrition for the world's people.[5] Results from initial work on stocking and growout production of channel catfish were presented at meetings and then published by Swingle in 1958. His findings, which showed that catfish could be produced commercially and economically, proved to be a significant stimulus to commercial farming of the fish.

Channel catfish had become a focus of full-scale research at Auburn in 1957. The number of ponds used exclusively for channel catfish experiments had increased that year from five to twenty-six. After Ellis Prather's success spawning the fish in 1958, Auburn was able to begin producing its own fingerlings, thus reducing its dependence on federal hatcheries such as Marion, which could not supply everyone's needs. Auburn researchers were able to move ahead with investigations of the management problems involved in raising catfish for food.

That same year, the FWS added catfish to the research agenda at Marion. Sending Kermit Sneed from Oklahoma to head up the program and hiring Harry Dupree as Sneed's assistant strengthened the facility's capacity for basic research on catfish. Dupree had done work identifying the vitamin and amino acid requirements of catfish. Sneed and Dupree continued advancing knowledge of nutritional requirements and methods of catfish culture at their new Alabama location. O. L. Green would later join the team as a trainee from Kentucky, where he had been a fish culturist for Sneed and Dupree at the FWS hatchery in Frankfort.

By 1959 Auburn was deep into work with catfish, and the program achieved an impetus of its own. The university was raising catfish in fifty-seven research ponds and had sold its first fingerlings. Over the next several years, they sold thousands of fingerlings until commercial producers in the state became established and were able to accommodate the demand. Information, experience, and technologies were on hand for successful production, and research results had shown it could be economically viable. By 1960 the ground was fertile for commercial catfish culture in the southeastern United States.

Notes

1. Until 1960 the official name of the university was Alabama Polytechnic Institute but for the sake of simplicity it is referred to throughout the book as Auburn University or simply Auburn.

2. Robert R. Stickney, *Aquaculture in the United States: A Historical Survey* (New York: John Wiley and Sons, 1996), 196.

3. Ibid., 197; Thomas L. Wellborn, "History of the Channel Catfish Industry," 12 February 1990, Special Collections and Archives, Auburn University, Auburn, Ala., 3; Frank B. Cross, comp., "Pondfish Culture in Kansas: A Selective Chronology," 1991, Special Collections and Archives, Auburn University, Auburn, Ala., 1–3.

4. Nick C. Parker, "History, Status, and Future of Aquaculture in the United States," *CRC Critical Reviews in Aquatic Sciences* 1(1) (1989): 97–109, citing J. T. Bowen, "A History of Fish Culture as Related to the Development of Fishery Programs," in *A Century of Fisheries in North America*, ed. N. G. Benson, Special Publication No. 7, American Fisheries Society, Washington, D.C., 1970, 71.

5. E. Wayne Shell, interview by author, tape recording, 16 April 1998, Special Collections and Archives, Auburn University, Auburn, Ala.

2
Getting Started

We didn't know what we were doing.

—Check Stephens and Richard True

The raindrops and wipers beat their steady rhythms on the car windows as Chester O. "Check" Stephens drove into the Gulf station in Greensboro, Alabama, that fall morning in 1960. He drove there, as he did every morning, to top off his gas tank for the day's work. Stephens was a Ralston-Purina feed salesman, a job he had held since graduating after World War II from Alabama Polytechnic Institute, known even then as Auburn University. As a salesman in a rural area, he often traveled long distances.

On that particular day, Paul Kieffer was riding with him. Kieffer was a veterinarian and researcher with Ralston-Purina in Saint Louis. He had arrived in Meridian, Mississippi, by train the evening before and had spent the night at the Stephenses' home in Greensboro. His mission on this trip was to find out why the U.S. Fish and Wildlife Service fish hatchery in nearby Marion was buying such large quantities of Purina Trout Chow from Stephens. He was particularly curious because trout was a coldwater species, and the Marion hatchery worked with warmwater fish. Kieffer had asked Stephens to take him to Marion so he could talk with hatchery director Jack Snow and get some answers.

The two men waited for the station attendant to fill the tank, and as they chatted, a truck drove up. It was Stephens's acquaintance Richard True. True was a cattle and cotton farmer in nearby Newbern who bought cattle feed from Stephens.

Stephens called out to him, "Hey, Rich! I'd like you to meet Dr. Paul Kieffer." True climbed out of his truck and walked over to greet Stephens and his guest. The men shook hands.

"What're you doin' today?" Stephens asked.

"Well," True declared, looking out at the steady drizzle. "I'm not really sure, what with it raining and all."

"Why don't you come with us to Marion?" Stephens suggested.

"I think I'll just do that," True responded after a moment. He left his truck parked at the Gulf station and climbed into Stephens's backseat. With the tank full of gas, the three men set off toward the hatchery in Marion.

On the drive they talked about this, that, and the other—and about fish. True decided to ask Kieffer for some advice.

"Paul, you're a fish expert. I've got a couple of fishponds down there by my house in Newbern that were stocked with fish. The mosquitoes in the ditch below the ponds were about to drive us crazy, so when they sprayed my cotton the last time from the air I had the boys spray down the ditch to kill the mosquitoes. It got in my fishing ponds and killed all my fish. That was about a year or so ago, and I'm ready to stock the ponds back now. What would you suggest I put in them?"

Without hesitating, Paul Kieffer responded, "I'd stock them with catfish, . . . yeah, catfish."

True considered it for a moment and said, "Well, now, that sounds like a good idea. Where do I get the catfish?"

That was the sticking point. "I don't know where you can get any fingerlings," Kieffer had to admit. "I don't know anybody who's hatching them." The men settled into the rest of their ride, talking about other things.

At Marion, they learned why the hatchery had been using so much Trout Chow. Trout Chow was the only feed commercially available for young fish at the time. Fish culturists there were raising bass fingerlings to a larger size for stocking in commercial waters to correct imbalances in fish populations and had found that the bass fingerlings would take the feed after they had been trained to do so.[1] The men finished their visit and drove back to Greensboro. On the way back, the conversation turned to catfish again, and Kieffer remarked, "If nobody's hatching them, well, there's an opportunity for someone."

Now Check Stephens had an enterprising spirit—a trait that served him well in his work as a salesman. He started thinking about Kieffer's idea, and the more he thought about it, the more he saw opportunity beckoning.

"Yes," he thought, "if nobody is hatching them, and you can't get them anywhere, that might be a real opportunity. *Somebody* needs to hatch them."

Stephens and True talked and talked. Clearly, there was a demand. A lot of the folks around did like catfish, and the catch from the river had declined in recent years. If there was a demand for them, whoever was hatching them could sell them and make some money. That clinched it.

"We'll give it a try," they decided.

They spoke to a local friend of True's by the name of Bryant Allen. Like True, Allen was a cotton and dairy cattle farmer, and he was looking for a way to make some additional income. He was intrigued by the possibilities they were discussing and agreed to join them in their search for information. Although he was farming at the time, he had been trained as an engineer, and all three agreed that with his knowledge and skill, they would be better equipped for the venture.

A crop duster's error, a visit by a curious feed company researcher, a fluke of the weather, a coincidental encounter at a gas station. . . . How could the three men, or anyone else for that matter, guess that these chance circumstances would play into the birth of an industry that would mushroom over the next forty or so years into one of the largest contributors to the state's economy and that of the entire southeastern United States?

Stephens, True, and Allen assessed their needs and their resources. True had two ponds and well water on his farm. They would need some brood fish and certainly some equipment and supplies. That meant they were going to need money. And if they were going to establish a business, they would need to think of a name to call their operation. First and foremost, they needed plenty of information and guidance, because they knew little about how to operate a fish hatchery.

Questions went around in their minds: How were they going to get brood fish? How would they get the brood fish to spawn? What would they do with the eggs once they were laid? What conditions would the eggs require to hatch and the fingerlings to grow? Could they provide those conditions on True's farm? What equipment would they need to do all this?

The men "went all over everywhere" looking for people who might be able to advise them about hatching catfish. They were able to obtain advice from the federal hatchery in Marion. "We got a lot of information from Jack Snow. Harry Dupree came in and helped us a good bit," recalls Stephens. Dupree and O. L. Green had just moved to Marion from Kentucky to work on Kermit Sneed's spawning project. Sneed and Howard Clemens had originally induced spawning in channel catfish at Oklahoma State University in the late 1940s. From Sneed, the three men learned facts about catfish that were not yet commonly known.

Stephens, having a bachelor's degree in agriculture from Auburn, knew about the fisheries program there. The men made the three-and-a-half-hour drive to Auburn more than once to talk with Homer Swingle and Ellis Prather. Prather demonstrated and explained to the men the spawning and hatching procedures he used. Bryant Allen paid special attention to the equipment in the hatchery shed at the research ponds, observing how it was made and how it was set up. With his understanding of mechanical processes and his talent for building machinery, he felt confident he could replicate what he saw.

"That won't be any problem. We can do that!" he told his two friends.

So they all agreed, "Well, let's go ahead and do it."

The Auburn researchers suggested the men visit Ben Nelson, a supervisor at Anderson Minnow Farm, a large baitfish farm in Lonoke, Arkansas. So they set off for Arkansas to see how farmers there were hatching catfish, hoping they might also be able to obtain some of Nelson's good quality catfish broodstock for their own use in the spring. The visit was very helpful.

In addition to the researchers at Auburn and at Marion, the men found an especially enthusiastic supporter in Earl Kennamer, who was at the time an extension agent with the Alabama Cooperative Extension Service. His expertise was primarily in wildlife matters, but he had been convinced for some time of the potential of channel catfish for commercial development and was eager to help the three entrepreneurs. His energy and support were invaluable to them, in Stephens's opinion. The men took a look at their finances and agreed to each scrounge up a thousand dollars to make a total investment of three thousand dollars. With that, they figured, they would have the wherewithal to prepare the ponds and purchase enough materials to build what they needed for a modest start.

The group needed a name under which to operate. While they sat around trying to find a name on which they all agreed, Stella True, Richard's wife, came up with one that sounded good to them. They would call their enterprise the STRAL Company, a composite of the beginning letters of their last names. A lawyer acquaintance of Stephens's in nearby Selma drew up papers, and their business became official, a "legal" entity. Now they could turn their attention to building a catfish hatchery.

At the same time, all three still had their regular work to do. Stephens was selling feeds for Purina. True and Allen both had their dairy businesses and crops to take care of. None could devote more than part-time attention and energy to the project. Yet things got done, on weekends and with an hour or two here and there during the week. Because of the fortuitous combination of their skills, the jobs broke down in a very workable fashion: Allen and True did most of the mechanical work, often in Allen's shop. According to True, Stephens was the "front man—a good salesman, good publicity man. We just all did what we could do best."

True's two ponds at his house in Newbern needed reworking, and the men reasoned that smaller ones would be more manageable and useful for starters. When the weather was dry enough, they drained the ponds, hired an earthmover, and divided one of them into two. The bulldozing was their biggest cost, according to True. "It was a local person with a 'dozer that we hired, who did that type of work. It wasn't a whole lot of building, because we just put a little levy in it—pushed dirt up. I believe we started working on the ponds during the winter, and by early spring they were done." The larger one was to be the holding pond for the brood fish. The two smaller were for fingerlings. The men planned to start using them early in 1961 for their first hatch. The well behind True's house overflowed into a holding pond, from which they connected a pipe leading to the ponds in order to fill them. True had a small dairy barn behind the house that would serve well as a spawning shed. The men cleaned it out and prepared it for spring. This was where the aquaria would go when it was time to start the spawning.

About fifty yards down the hill from the barn, the men cleared a rectangular area of about three hundred square feet next to the ponds. They sank creosote-treated posts around the outside of the clearing and put siding up around three sides. The fourth side they left open. With a corrugated metal roof, they created a space that was protected yet well ventilated and shaded from the hot Alabama summer sun. This would be the hatchery building.

When the STRAL partners had observed the spawning setup in Arkansas the previous fall, Ben Nelson's aquaria had been empty. Fall was not spawning season. But they had been able to examine the aquaria and the other equipment he used. "We decided to build twenty aquaria," True recalls. "We started in the fall, built the aquaria, built some of the equipment we needed, and fixed the barn where we could put it in and be ready for spring. . . . We used three-quarter-inch angle iron—cut it, welded it together, and then used a heavy, quarter-inch glass and just regular aquarium cement. The bottom of the aquarium was green plywood, and the four sides were glass. They were about two feet square and about ten inches deep."

The men had been told that the aquaria would need to have water running through them continuously in order for the fish to spawn. But they had no generator and did not have enough money to buy one. So they used a simple water pump. They had learned also that they were going to need some hatching troughs. Nothing they already had fit that bill. They had to build them and copied basically what they had seen in Arkansas. Since Nelson's had been wooden, they built wooden ones. They estimate the total cost was less than a thousand dollars.

The troughs they had seen had had wire baskets hanging in them in which the spawns, or egg masses, were placed for hatching. The three men built their own out of quarter-inch hardware cloth—wire screening—that came in about one-yard widths. They cut and folded the cloth into a shape that would hold the egg masses until the freshly hatched fry dropped out.

After about a week, when all the fry had absorbed their yolk sacs, they would need to be transferred from the troughs to holding tanks. The men had nothing they could use for that purpose and could not afford to have anything made. After looking around and thinking a lot, they hit on the idea of using ordinary concrete burial vaults and promptly bought four. Those were affordable containers to hold the little fish.

The men needed agitators to keep the water moving in the troughs and holding vats. An acquaintance who worked at Alabama Power Company in Greensboro brought the three entrepreneurs in touch with a company engineer, Junius Kendrick. Kendrick told them what type of gear motor they should use, one that would have the proper revolutions to turn the paddle wheels that were to move the water in the hatching troughs.

After a state biologist came to check the pond and the water and gave his official "stamp of approval," the friends were ready to collect fish for broodstock. The three men had bought some brood fish from Ben Nelson. They received some from AU and also went to local fishermen who fished the nearby rivers. The partners asked the fishermen if they would save channel catfish they caught in their normal routine of fishing, ones that were between one and five pounds, True recalls. "They were commercial fishermen, catching and selling fish to the general public. Most of the [fish] came from the Warrior River, right here. We paid them the going price. We had several we knew here in Greensboro, and they were happy to do it for us."

The most exciting experience for the men came, however, when they went out to catch some of their first brood fish themselves. Older residents, they knew, had been fishing the river for catfish as far back as their memories reached. "The old timers, they'd go out what they call 'grabbing,'" says Stephens, describing how it used to be, and sometimes still is, done. "The catfish male finds a place to spawn, then he goes and gets a female, and they go into the hole to spawn . . . maybe a hole in the bank. . . . At night, the fishermen will walk along the edge of the river, and they'll find a hole and reach up in there and get the catfish. *It'll* probably grab *their* hand. You could stick your whole hand in and pull them out, and that's what they call 'grabbing.' I didn't ever get up enough nerve to do that. But there are a lot of people that do grabbing now. And they aren't scared of a snake . . . a snake can stay up under there and live, you know."

With equipment from local fishermen and the knowledge that others had done it before, the three adventurers set out to the Alabama River and to the Tombigbee. They ran the fish baskets at night. "We'd take these boxes you put in the river to trap fish in," Stephens tells. "We made some of them. Some of the river fishermen that had boxes, we'd buy some of the boxes from them. . . . You put cottonseed cake in there, cottonseed meal, compressed. And you put them in the river. You'd maybe have a float on them, a bottle or something where you could tell where they were. You'd usually run your boxes at night—go down at night and take them out. Everybody thought the catfish fed better at night. We put some next to the bank and some of them out in the river, just sink them and tie them onto something. And that's where we got them. It was a lot of fun . . . and kind of scary now and then, 'cause you'd pull out snakes in the boxes. They'd be in there to eat the fish."

The partners managed to pull up catfish and brought them back to their brood pond. There, the fish would stay loose in the pond until spawning time. They carried the fish in burlap bags, unlike how the fishermen did it, recalls Stephens: "We got some croaker sacks, and we put the fish in the bag and took the fish and put them in the fish tanks [on the truck]. Sometimes we'd put more than one to a bag." By using some already established techniques and some they devised themselves, the three partners managed over the weeks to collect between fifty and sixty brood fish for that first year—twenty-five or thirty females and a comparable number of males.

None of the three men knew what to feed the fish. The freshly rebuilt ponds did not offer enough natural sources of the nutrition they would need. Purina researchers suggested Hog Chow. When they consulted Ellis Prather at Auburn University (AU), he suggested they try hog liver. A slaughterhouse and meatpacking company that dealt with livestock in Tuscaloosa, Ziegler Packing Company, had large amounts of condemned beef and some hog liver. The men went there and bought livers in fifty-five-gallon barrels. They cut it up at the hatchery, put the pieces in bags, and froze them in an old freezer. They fed beef and some hog liver to their first broodstock to provide what the fish needed for better spawning.

By May 1961, when spring had come to Hale County, the STRAL brood fishpond held all the adult fish the three men felt they would need. The fingerling pond was full of water, ready to receive the young fish. The hatching troughs and holding vats stood waiting in the three-sided hatchery. Approximately ten or twelve aquaria sat lined up in the small spawning barn behind Richard True's house. The time had come to see whether all their work would amount to anything. It was time to attempt their first spawn.

The task facing the men initially was to catch the fish and choose the ones they wanted to use. An acquaintance who lived close by, Dave Thornton, provided a fourth set of hands to help with that and with some of the other work. They pulled a seine, a large net held down on one side by sinkers and held up on the other side by floats, through the brood fishpond. By trial and error, the men gathered the fish together to one side, where they would be able to sort them. Together, they examined the fish carefully, one by one, to see how their bodies were shaped and how large they were. If they were large and had what looked like a good "catfish" shape, and, in the case of the females, if they looked as though they were "ripe" with eggs, the men would use them. The others were returned to the pond.

The partners had learned how to sex catfish by watching Prather handle fish at the AU research ponds and by talking to old-timer fishermen and people like Earl Kennamer, Kermit Sneed, and O. L. Green. Now, however, they were on their own.

"You have to catch the fish by the tail, between where the tail fins [come] out, and squeeze—it paralyzes them—and make them immobile," says Stephens. True was responsible for most of the sexing, since the fish and the spawning shed were right by his house. He chuckles when he recalls that first experience: "That was a learning process. Once you know what you are doing, it's very simple. But it is difficult to get on to. I am sure we paired some females together." That might explain why not all the pairs the men selected produced a spawn.

Before they could place the females into the aquaria with the males, however, the men had to inject them with chorionic gonadotropin, a pituitary hormone that would make them lay their eggs. This was what Ben Nelson had used under Kermit Sneed's guidance. "It was a veterinary product derived from horse urine. They told us the dosages they had tried. We tried different dosages ourselves. We really didn't know. It was pretty much trial and error," explains True. The three men caught and identified the females, filled the syringes, quickly stuck the hypodermic needle into the female's abdomen, and injected the hormone.

Finally, all the aquaria were full with breeding pairs of large fish. The fish could barely turn around. Now, the men could only hope, wait, and watch. If they saw no results within twenty-four hours, their instructions were to repeat the procedure. Within a week to ten days, they would know whether or not a pair would spawn. They were to keep the fish confined there no longer than ten days. If a pair did not spawn, they would put the fish back in the pond and try another pair.

~

Like excited, anxious fathers-to-be, they worried about things that could go wrong. They had spent months of thought, time, energy, and several thousand dollars of their own money. This was uncharted territory for them. Things really could go wrong. Electrical outages were common, and if the electricity went off, the pump would quit, and water would stop circulating through the aquaria. The big fish in those relatively small containers would run out of oxygen in a matter of minutes and die. Keeping them in a proper environment was critical.

There were other reasons, also, for keeping a close eye on the fish. When a female catfish finishes spawning in nature, she leaves the spawning site. The male hatches the eggs and takes care of the young. If she does not leave, the male will chase her off. In this instance, however, the female was trapped in the aquarium and could not leave. The partners found that if they did not remove her after the pair finished spawning, the male would kill her. In such close quarters, the aggressive maneuvers of the male would also tear up the eggs. So someone had to be present to take the female out quickly and put her back safely into the pond.

The three men would have to learn many things the hard way. Not realizing, at first, just how closely they had to keep watch, they went to check on the fish several times only to find mangled, dead females and torn-up egg masses. That was when they realized they would have to be there basically twenty-four hours a day.

The men took turns keeping watch around the clock. Because the little spawning barn was only yards from his back door, most of the night-watching fell to Richard True. The first time he went out after dark, he was surprised to see half the fish on the floor. "When I turned on the light, they began to move around," he recalls. "In the daytime they were doing great.

At night, half of them jumped out." With all the water streaming through the aquaria, there was "quite a sheet of water on the floor. We realized we had to make some kind of screen to put on top and weight it down so they could not get out. We caught them in time. We probably did some damage stress-wise, because we couldn't determine whether that kept some of them from spawning."

True did not get much rest during that first spawning season in May and June. He would lie down and set the alarm clock for thirty minutes, and when it rang, he would get up to go check on the fish. Then he would lie down again and set it for thirty minutes later. Before the season was over, he was sleeping on a cot in the barn, with an alarm clock.

Stephens remembers the wonder of the first spawning. It was "really, really, something," he says. "You go down there, and you see those fish swimming around in those aquaria, see the males and the females. They would swim around. If you'd watch close, you could actually see where they would just kind of embrace. It'd be head-to-tail, and you could see the eggs come out. About the time the eggs came out, the sperm would fertilize the eggs, and you could watch that egg mass just build up. And he'd swim all around, kind of roll them up."

Once a female catfish was back in the pond, the eggs needed to be taken out so another female could be injected and placed in with the male. Removing the eggs was at first easier said than done, however. The men put their hands into the water to pull the eggs out and found the eggs stuck to the wooden bottom. As they worked with their hands to undo the eggs, they upset the male, who would then attack the intruding hands and, in his distressed state, destroy the eggs. When they did manage to get the eggs out, the bright yellow mass came out so damaged that many of the eggs were no longer protected by the gelatinous coating in which they had been embedded. There had to be a better way.

Pooling the ingenuity on which they would draw many times, the three came up with the idea of placing something on the aquarium floor *before* the fish were added—something to which the eggs could attach but which could be quietly slipped out without upsetting the male. "What would do the trick?" they wondered. Would tar paper work? They cut a piece of roofing paper to fit the aquarium bottom and lay it down under the water. It worked like a dream. The egg mass stuck to the tar paper, and slipping their hands gradually in under the male, the men could grasp the sheet by an edge and ease it up and out. The fish did not seem to notice what was

Fig. 3. Richard True and Check Stephens at the STRAL hatchery, 1969.
Looking at an egg mass at a STRAL Company hatchery trough. *Left to right*: Richard True, Hale County extension farm agent Lee Gober, and Check Stephens. At the time, STRAL Processing Company was anticipating going into business supplying Miz Diz Catfish Shanty restaurants. (Photo courtesy of the *Selma-Times Journal.*)

happening; the men's hands and the egg mass were spared. Success! All that remained to do was loosen the spawns from the tar paper and take them to the hatchery. From then on, tar paper rectangles became a staple in their spawning inventory.

The men carried the eggs down to the hatchery in buckets and put the whole spawns into the wire baskets they had built (see fig. 3). They left the baskets in the troughs. The paddle wheels turned in the troughs, moving the water up and down, gently lifting and lowering the egg masses the way the male catfish's tail and body would do in nature.

The hatchery troughs may have looked like a safe haven for the spawns, and the men may have thought their sleep-deprived nights were over. Not so. A twenty-four-hour watch was needed here, too. If the electricity went off, the paddle wheel would stop turning, and the water would stop lifting and lowering the eggs. If the pump bringing the water from the well pond failed, the troughs would not have the running water needed to get oxygen to the eggs and hatching fry.

The three entrepreneurs repeated the entire process as long as they were able to obtain spawns. They needed as many eggs as they could get. Not all the fish spawned at the same time, so the tension and sleeplessness of spawning season lasted through May and June and into July. Altogether, perhaps 10 percent of the brood fish pairs produced spawns. Through trial and error, the men learned about the importance of adequate aeration. They lost a few brood fish because of their own lack of experience and knowledge. Some of the eggs were destroyed by fungus. Often, the three did not know anything was wrong until a mass had already deteriorated.

May and June 1961: The paddle wheels turned slowly and steadily between the homemade wire baskets in the troughs. Bright yellow egg masses rose and fell with the gentle current. In spite of their lack of sleep from working tirelessly for weeks, the three men bent over the vats and watched, curious and amazed. As days passed, eggs began to turn red with the blood forming in them. Vague, tiny forms began to take shape. Little tails and two tiny black eyes appeared. By four to ten days after spawning, depending on the water temperature, the "sac" fry had all dropped out of the gelatinous masses down through the bottoms of the baskets to the floor of the trough, where they stayed—at first. Within several days, the "swim-up" fry started coming up from the bottom and began moving around. The men were energized.

As it turned out, that first hatching was a near disaster. The men had built the troughs out of wood, which was a mistake, according to Check Stephens, "because they had little cracks in them where we put the boards together." Before long, the partners noticed tiny fish flopping around on the floor of the hatchery shed. "We lost some," says Stephens. "The fry were so tiny they could slip almost through anything." The men could do little to prevent more fry from slipping out. They could only try to save as many as they could and chalk the loss up to experience. Placing the end of a hose on the trough bottom, the men siphoned the remaining fry into big washtubs and then lifted the heavy tubs, carrying two at a time between the three of them, to the fingerling pond. The work was backbreaking, but the men managed to save some of the young fish.

By the middle of July, all the fry they could transfer with the washtubs were swimming in one fingerling pond. Now the men had to feed the tiny fish. There was no feed manufactured for catfish at the time, much less for catfish fry. Stephens, who was still working with Ralston-Purina, talked with people at the company's main office. The Purina researchers suggested the men use Hog Chow, a feed for hogs that was about one-third protein. The tiny fish were much too small to get the Hog Chow pellets into their mouths, so the men ground the pellets up. Over the weeks, they gave the fry a variety of feeds, depending on what they could obtain at the time. Sometimes it was Hog Chow, sometimes Trout Chow, and sometimes another feed.

As the STRAL fingerlings grew over July and August, the men began to throw whole pellets out over the pond. When they tossed the pelletized Hog Chow onto the water, it sank to the bottom. They had no idea at first whether the fish were taking the feeds or not. They could only hope that the fingerlings were eating it off the bottom, because no one could see them eating. Trout Chow, in contrast, floated on the surface. When they used that feed, they noticed after a while that the young fish were coming to the surface to take it.

By September the men were able to make out young catfish in the pond water, especially around the edges. Now they needed to find people who would buy them. Richard True and Bryant Allen had to take care of their farms and could not travel much. Stephens, on the other hand, was already going on sales trips around Alabama and other southern states as part of his job, and he began to use those trips to talk to everyone and anyone who would listen about STRAL's first crop of catfish fingerlings.

Stephens had the perfect combination of skills, knowledge, and experience needed to sell and was a natural promoter for STRAL. He put the connections he had already established through his feed sales to good use for the hatchery. He lost no opportunity when driving through towns and county seats: "I'd stop and go in and see the Soil Conservation people, who were the ones who laid out the fishponds. There's probably not a county or a county seat in Alabama, Tennessee, Georgia, Mississippi that one time or another I hadn't been in some several times. That's the way I spent a lot of time. You didn't have any problem talking to the people about fish."

There was a great deal of curiosity about this new thing called a catfish hatchery. The three partners had green and white business cards that featured a drawing of a catfish. The STRAL card opened doors for Stephens, doors to which even his Ralston-Purina card failed to gain access. When he wanted to talk to people about feeds, he would give them his Purina card. Often their reaction was barely a "so?" However, when he told people he was from STRAL Company catfish hatchery, they asked right away, "Do you have a card?" Many times, he would give the receptionist the card with the fish on it, and within seconds, a man would come out of the office to see him, wanting to know what this new thing called a "catfish hatchery" was all about. "Come on back into my office and tell me about it," the man would say.

"Everybody wanted to talk about it," recalls Stephens. A dentist he went in to see had "somebody in his dental chair with their mouth open, wired up" and came out to talk with him. "One day he even went with me to a fishpond, and [his patient] in there like that. That is how much effect this new industry had."

STRAL's first ad read something like: "Channel catfish fingerlings. Write for prices. We deliver." A phone number followed. The telephone response was surprising, and it was a challenge to keep up with the names and addresses of interested people. "Every time we got [a call], I'd card it," notes Stephens. "I'd do all that work at night. Sometimes it'd be midnight before I'd get through, just a day of carding them, writing them, and sending them a price list. And then we had a booking!"

Note

1. Jack Snow, telephone conversation with author, 20 August 2004.

3
In the Fingerling Business

We were just keeping our heads above water.

—Check Stephens

September 1961: Stephens got an order for about five thousand fingerlings from a tire salesman who had a fishpond. His name was Billy Hicks, and he lived in Luverne, not quite a hundred miles southeast of Newbern as the crow flies. To get the fish to him, the STRAL operators had to first get the fingerlings out of the pond. Dragging a seine through the water, they managed to get a few fingerlings up, but far too few for the sale. They decided to drain the pond to better catch whatever fish might be there.

The sun was coming up early the next morning when they began to let some of the water out of the pond. By the time the water was low enough to pull the seine through again, the air was hot and heavy. O. L. Green, who worked at the Marion fish hatchery and had given them advice before, came driving by. He stopped his car, jumped out, leaving the door open behind him, and ran over to them. "Hey! Y'all are gonna kill those fish!" he shouted to the men. "It's too hot! Put some more water in there!"

The men did as Green said. They turned the well pump on and added water back into the pond. They had not realized that the fish would not survive the stress of the September heat. "O. L. saved us that first crop of fingerlings. We almost lost them. We didn't know any better," recalls Stephens, shaking his head.

Now they had to count out five thousand small, slippery, wriggling fish and keep them alive and well in the process. The men built a glass-top counting table. It was "like a pool table—little sides around it—and it sloped slightly to the end," recalls True. "I think it was about two-and-a-half feet wide and about four feet long," remembers Stephens. "On one corner, we'd cut a little off the glass where we could pull a fish up, then drop it through into a bucket." Stephens recalls that he, True, and Allen "would sit there at night and count those fish until we got to five thousand." For the first year or two, they counted out every order in this way. "You get where you can do it pretty good," True chuckles.

Allen had an old, beat-up pickup truck that he and True had adapted for deliveries. With marine plywood, they had built a tank to haul fish in. According to True, they had "put in aeration, a four-by-six-foot tank, with four compartments, to fit in the back. We just guessed, because we had never hauled before. But we had seen hauling tanks. The state had tanks it hauled bream and bass in, so we knew, generally speaking, there was a way to haul fish." The tanks had a hinged lid, made out of the same plywood, and were portable, so they could be mounted on the back of any pickup truck. The men mounted them on the back of Allen's truck and placed agitators inside to provide oxygen for the fish. The agitators were connected to the truck's battery. A flag on the top of the agitator shafts showed whether the shafts were turning sufficiently.

When the time came to deliver the fish to Billy Hicks in Luverne, the men turned on the agitators and poured the fish in. Allen and Stephens then took the truck to Hicks's pond and emptied the fingerlings out into the water. They watched with relief as the young catfish swam off. Their first delivery was a success. They got their money, and everybody was happy. They drove home, pleased as punch. They were really in the fish business!

Customers expected, naturally, that fingerlings they had ordered would arrive at their ponds in good condition. However, transporting the fish was tricky. The jolting ride inside the small tanks was stressful for the fish. It could be stressful for the fledgling hatchery operators too, particularly when they had to deliver fingerlings very long distances. STRAL's first out-of-state delivery trip was a nightmare, True says. "We delivered them to Saint Louis, Missouri, and we didn't know what we were doing. Check found a friend of his, a Purina salesman in Saint Louis, who had a big fish-out lake up there, and he talked him into buying our fish."

"Are we all set?" Allen asked True late one fall afternoon in 1961, as they made sure everything on the back of the truck was secure.

"Yep," answered True. "Got the map and address?"

Allen brushed his hands off and opened the cab door. "Sure do."

"Well, let's head on out," responded True, as he climbed in on the other side.

The men eased the old truck cautiously over the dirt road near the edge of the fingerling pond, bouncing and swaying gently out to County Road 16, which ran in front of True's house. They were on their way. Checking through the rear window of the truck, they could see the aerators' flags turning. "Looks good," said True. They were satisfied.

Saint Louis is about six hundred miles from Newbern. The men planned to take turns driving all night in order to deliver the fish during daylight hours. Before they left, they had gone over in their minds possible problems that might come up during the trip. They missed one.

When twilight fell, they turned the headlights on. Thirty minutes later, the lights on the dashboard and the beams on the road ahead began to dim. True turned around to check the aerator shafts. In the darkness, with the help of light from a passing car, he made out just enough to tell that they were barely turning. Allen slowed the truck and brought it to a stop at the side of the highway. The flags didn't speed up. The two men discussed the situation. The battery, which had powered the aerators during the daytime, must not have been able to carry the load of the lights and the aerators at the same time.

"What are we going to do?" the two men asked themselves. They couldn't turn the truck around and head back to Newbern. The only thing they could do, they decided, was to continue traveling and use the lights sparingly. They turned back onto the highway and picked up speed. When they could slip behind a big truck, they cut off their lights and navigated by what they could see of the road lit by the beams in front of the truck ahead of them. When they came to a town, they turned the pickup lights on long enough to get through without drawing attention and getting a ticket. Neither of them got much sleep that night.

"It was a mess," recalls True. "We had our ten thousand little fish, and at that time we probably were getting twenty cents for them. So a couple of thousand dollars we had on the truck, and that was big time for us. We didn't want to lose it."

Once the men made it to Saint Louis safely, they proceeded to fumble through stocking the lake: "We had been nursing those fish all the way up there," says True. "But we didn't know much about putting fish in the pond. . . . We were trying to be real professional with our first hauling. We had plastic buckets, and we filled one half-full of water and dipped up some. I started down the hill with the bucket, and the bail broke. And there were the [fish], right there on the bank, on the ground. Best we could—we told [him] that it wouldn't hurt them—we picked them up and threw them in the lake. We probably didn't get any survivals. We never heard. It was a big lake, and he was just stocking a few for fishout. So he probably never knew."

The first crop of fingerlings in 1961 amounted to a little less than fifty thousand fish, so few that they barely needed one-half of the divided pond on True's property. After all the fingerlings were sold off, the men were pleased. The year had been difficult. But then, all three of them were already accustomed to working from daylight to dark whenever something needed to be done, no matter what it was. Although they had never known just when a check was going to come in, they had made a little money, enough to pay for some of their costs. A lot of problems had been solved, and the men could see that catfish had a big future. And as far as they could tell, they were the only ones doing it.

"You're crazy," the bank officer declared with a friendly chuckle. "Why, the Warrior River is full of catfish for anyone who wants them. There are more in there than people will ever eat. And you think you're going to go sell them when folks can go get them for nothing? That's just a bunch of dreams!"

Optimistic after the success of the first year, the STRAL entrepreneurs had decided to ask for a loan of around two thousand dollars. They were essentially laughed out of the office when they approached a local bank in Greensboro early in 1962. The notion of making a profitable business out of spawning catfish for people to grow in ponds seemed ludicrous to many people in those days.

The three did not give up, however. They tried People's Bank and Trust in Selma, where Check Stephens had an account. Stephens told the loan officers about their success so far, their plans for the hatchery, and how much money they thought they would need to finance a second year of operation. Whether or not people at the bank perceived some potential in raising catfish, they agreed to lend the men six thousand dollars. With a bank draft for three times more than they had originally requested, the men were so pleased that they didn't mind the shaking heads and smiles that followed them as they walked out of the building.

They set about making needed changes for producing a second crop. The experience of the first year had given them ideas for improving their operation. The information they had received from others was helpful but in many ways incomplete, and so much of what they learned about operating a catfish hatchery came through trial and error. Often, they had to put their heads together on the spot to find a solution to a problem they had not foreseen. More often than not, they found out whether equipment they built worked or did not work only after they began using it.

By spawning season the next spring, they had built two additional hatching troughs and had metal liners made for all four troughs. That way, the newly hatched fry would not escape the way they had the first time. They kept the tiny fry in the troughs longer than before, from ten days to two weeks. That allowed the fish to grow somewhat before they were put into the less-controlled environment of the open pond. Anticipating a larger fingerling crop, the men had a few more ponds excavated on the land behind True's house.

Spawning and hatching went more smoothly and successfully this time. When it came time to deliver the fingerlings in the fall, the men made sure to harvest the fish at night or during the cooler hours of the day. With the loan money, they were able to buy a truck with a heavy-duty alternator that could keep vehicle lights and aerators for the tanks going at the same time.

Crickets were serenading each other and the three busy men near the edge of the pond one warm and humid fall night. The bare light of the bulbs hanging from the hatchery ceiling illuminated the men and the small table over which they were bent. It could have been 1962, or perhaps it was 1963. There were many nights like that one. The air had cooled somewhat, now that the sun had been down for five or six hours, but sweat still ran down their faces as they emptied buckets of small, slippery fish onto the shiny counting table. With their hands, the men swept the fish, by twos or threes and sometimes by fives, across the wet glass and let them drop through the corner hole into the bucket waiting underneath.

It was two o'clock in the morning, and the men were counting out fingerlings they would be taking to Georgia in a few hours. They swatted their necks, faces, and arms when mosquitoes bit them. Two wires were strung along the side of the table. On one hung some metal washers. On the other, some square nuts. When one of the men had counted to one hundred, they moved one of the washers on the wire to the other end. When they had reached one thousand, they would slide one of the square nuts hanging on the other wire to the opposite end and return the washers to their original position. This makeshift abacus was one of the ways they used to keep track of the fingerlings they were counting.

They had been at it a good while already—it was a large order—when

Richard True looked up from studying the nuts on the wire and asked, "Did I move that nut over the last time or not?"

Stephens let the last little group of fingerlings wriggling under his fingers slide down through the hole into the bucket, straightened his back, and shrugged his shoulders.

"Gosh, I don't know," he said, baffled.

"Well, if I didn't, the customer's getting a thousand fish too many. If I move it now, he might get a thousand too few," said True.

The men wiped their glistening foreheads with the backs of their wet hands and paused to consider the situation. They came to the same conclusion: better give the man a thousand too many than a thousand too few. They resumed their middle-of-the-night task.

Orders were coming from all around, and their business was growing steadily. Clearly, the men could not go on using such a time-consuming and inefficient method. "We figured there's got to be a better way to count that'd be more accurate," Stephens explains. They applied a method they had heard about from experts at AU who used it for counting tiny, newly hatched fry. Other fish-farming industries were already in existence at the time, and they were probably already using it. They counted 250 fish and weighed them, then counted 250 more and weighed those, says True. "And if they came right out on the money, we knew how many fish were in a pound, or how many pounds it takes to make a thousand." With this information, they knew that if a hundred fingerlings weighed a certain number of ounces, harvesting a thousand weighed ten times that. The industry still uses this method of estimation today.

Why did they count the fingerlings at night? The men had learned that the young fish could not be handled safely in the sun or heat of Alabama's summer. The fish would require less oxygen and experience less stress if counting were done in the cool of night. In addition, the men preferred to leave with the delivery truck just before or around daybreak, so that they could do most of their traveling before the day heated up. By afternoon, the trip became more stressful for the fish in the small, unrefrigerated tanks.

~

In those days, researchers and farmers generally assumed that catfish did not eat during cold weather, so they did not expect the fish to grow much during the winter season. Thinking feed would just be wasted, therefore, few people fed their fish during the coldest months of the year. The hatchery operators observed differently, however, from their own experience with

Milton Bowman, a feed researcher with Ralph Wells Company in Monmouth, Illinois. Bowman had purchased a large order of older fingerlings, and the partners drove to Illinois to stock them in small coal strip-mine lakes. They were going to do feeding experiments through the winter, explains Stephens. "Richard and I left with two trucks and spent a day putting out all those fish." The Ralph Wells researchers "counted every one of 'em, had accurate records. The experiments turned out real good. They fed the fish all winter long. Bowman said, 'I know they'll feed, because sometimes we'll cut through eight inches of ice in order to get the feed to 'em.' And they fed."

One year, possibly 1963, the partners began to feed their own fingerlings through the winter. When seining up the fingerlings the following spring, they realized that the little fish were larger and stronger and that more had survived than had in earlier crops that had not been fed through the cold weather. "We just kept throwing feed in there for them," notes Stephens. "It was still a sinking feed. [They were] still feeding on the bottom. Nobody could see them. But those fingerlings doubled in size in about three months' time, by weight. They were eating. And the survival rate was much better."

The U.S. Soil Conservation Service (SCS) sent Elvin Wright to Greensboro in 1962 to work with the Watershed Association and to plan and apply conservation practices for the farmers of Hale County. Wright was an engineer who advised Hale County residents on the construction of their ponds. He got to know the STRAL partners and helped them when they built additional ponds for the growing hatchery. He was intrigued with this new way of obtaining catfish and talked about it to all the pond owners he met on his job. More ponds meant more potential customers for STRAL fingerlings.

At the beginning STRAL took every order they could get, no matter how small. On his road trips, Stephens would spend time in county extension offices and, with the extension agent's help, obtain orders from twenty-five different people in a day, people who wanted to have a little fun with raising catfish and ordered five hundred to a thousand fish. Even people with small ponds were deciding to give "this catfish thing" a try to see what might happen.

Rather like the FWS hatchery truck delivering bream and bass, the partners would meet the customers at an agreed-upon point such as a country store, a feed store, or even a courthouse. They would put five hundred or a thousand three- or four-inch fingerlings into a plastic bag with some water, blow oxygen into the bag from tanks they carried on the truck, and receive their money in exchange for the fish there at the side of the truck. They would get up and load the truck at four o'clock in the morning and spend all day delivering thirty thousand fingerlings in that manner.

For the first couple of years, True and Allen did most of the deliveries. They sold fingerlings starting in late July or in August. As their production grew and orders increased, sales continued through the winter, some years right up until the next year's hatching was about to begin. Usually, though, by the time hatching season came, all the previous year's fish were sold.

Truckloads were delivered all over the Southeast, "from the Tennessee line to the northern part of the Florida panhandle, from the Mississippi line into Georgia, around Cordele, Americus, in the south of Atlanta."[1] The distances were long, and there was little time for resting along the way. Stephens remembers trips to North Carolina in their large, two-and-one-half-ton truck. "One year we'd taken a load of six- to eight-inch fingerlings up there," he says. "Stocked them all over. It took us about forty-eight hours to do it, and I've never been so tired in all my life."

Mississippi did not turn out to be much of a market for the Alabama-raised fingerlings. The STRAL operators did stock some ponds in eastern Mississippi. In 1962 and 1963 STRAL stocked fingerlings in ponds Ruben Prescott had built to supply his fish restaurant in Columbus. When they grew large enough, Prescott harvested, processed, cooked, and served them to the public. Ruben's Fish House was one of a number of catfish restaurants that sold fish grown out from STRAL fingerlings during the 1960s.

STRAL's very first shipment of fingerlings had gone to Billy Hicks in Luverne. The next year, Hicks shared his first harvest inadvertently with the residents of the town. He had been feeding them Hog Chow, which sank to the bottom of the pond. As a result, he had not been able to see the fish feed. Hicks concluded eventually that he had no fish and decided to drain the pond and restock with more fingerlings. "So he pulls the plug on 'em," says Stephens. "Lo and behold, I think 120 percent of them had lived! There were tons of them, all over everywhere. He didn't know what in the world to do. Couldn't put water back in the pond. So he called everybody in Luverne to come out and get fish. Some of them might have paid him, but they got fish out of him."

People heard about STRAL from their neighbors and acquaintances and from the newspaper articles that were written about the hatchery. The residents of Newbern and Greensboro watched the new enterprise with a great deal of curiosity. They came to see what the three entrepreneurs were doing, and they were fascinated. It was odd . . . completely different . . . something new.

These were small communities in which almost everyone had either direct or indirect knowledge of what other members of the community were up to, and whether they did or not, they certainly talked about it. Stephens still chuckles when he thinks about how some people reacted to the goings-on at the hatchery during one of their first spawning seasons. "There were a couple of old maids who found out that we had those fish down there, spawning in those aquaria," he remembers. "They came down there, and they sat and watched and got the biggest kick in the world watching those fish spawn. That was one of the highlights of the year, because everybody was real interested at what these old maids were doing down there, watching our fish spawn."

The news spread outside the local communities and outside the state as well. By STRAL's second year, people from other states were paying visits to observe the hatchery operation. Businessmen from Mississippi came during the 1962 spawning season. They came to learn from STRAL's operation just as Stephens, True, and Allen had learned from Ben Nelson in Arkansas two years before.

In those early days of commercial production, many were comparing the catfish industry to the poultry industry, which had already developed a strong foothold in the South. Richard True remembers how the Mississippi visitors "thought this might be the beginning of something big. I do believe at least one of them was very big in the poultry industry at that time. I suppose we put on a big front for them. We didn't know much, but we knew more than *they* did." The Mississippi visitors went home and within a year had bought a tract of land and built many ponds. They built a hatchery, began to produce fingerlings, and had great success. A large catfish industry developed in Mississippi after the visits to that small hatchery in Newbern.

~

Much of the Mississippi Delta land was already covered with ponds. A number of farmers there had tried raising buffalo fish, Stephens recalls. "They had taken those rice paddies and put fish in them. Even before the catfish, there were bass and bream ponds. Everybody had a little farm pond somewhere." By STRAL's second or third year, Stephens and True were invited by a farm supply salesman in Mississippi by the name of Ira Bradshaw to talk to Mississippi farmers about raising catfish. The first of a number of group meetings was to be held in Laurel, Mississippi. "The primary purpose of us being at that meeting was to provide a market for the fingerlings we hatched," Stephens recalls.

The two men drove over to Laurel, discussing their plans for the presentation. Bradshaw had invited farmers from the area to gather at the county courthouse. When he caught sight of them entering the room, he strode over to welcome the two men and shook their hands. Stephens and True then began to prepare for their talk.

Word of the presentation had spread like wildfire among the farmers. The size of the group attending exceeded everyone's expectations. The room was packed with people greeting and talking with one another. They were curious and eager to see what these two Alabamians had to show them. Not a chair was empty. People sat on the window ledges, legs dangling against the courtroom walls. Those willing to stand stood in the aisles on each side or in the back of the room. Among the audience, Stephens recognized county agents as well as Mississippi farmers and salesmen he had met before. Some of the individuals there later established catfish farms that are still in operation today.

The partners could not have asked for a more receptive audience, and they were primed for their pitch. They had taken photographs of their operation and showed slides of the entire process, including spawning and hatching. They talked to the group about what they knew best: "We had pictures of our ponds while they were under construction . . . showed them how they were laid out. Showed how the dams were built, how to put the drains in. Showed the seines, how to seine them, how to get them out—the big fish. Showed the brood stock. Showed pictures of Richard holding the catfish, my hands full of little fingerlings."

The audience was spellbound, as was everyone when they saw the catfish spawn or the fingerlings for the first time. The men impressed their listeners with the potential that lay in catfish farming. That meeting and others like it gave STRAL's business a huge boost.

~

When Stephens was not with True or Allen delivering fingerlings, he was making contacts to sell more fish. His many years of experience selling feeds served him well but never compared with selling catfish fingerlings. Interest was that high. Calls streamed in, often inquiring as to the size of fish they were selling and whether the hatchery operators would visit ponds and advise farmers on how many fingerlings to stock per acre. Naturally, they also wanted to know what the fingerlings would cost. Stephens recalls,

"Almost everything we sold was a two- to three-inch fingerling for a nickel. We would deliver for ten cents a mile, one way."

Not until the fall of 1962 could the hatchery provide any appreciable amount of fingerlings for delivery. They hatched and sold in the neighborhood of one hundred thousand that year. Stephens sold fish from North Alabama to Florida, from the Mississippi line to the end of Georgia.

The money that came in was put right back into the business. There was no income to speak of for the first couple of years. "It was real bare bones," recalls Stephens, "because we took no money out of it. All of us had to get whatever money we had [from] what else we were doing." In a tape recording made for Wayne Shell at Auburn, Stephens elaborates on the early days. "We were taking about a hundred dollars a month out of the business. We were just about starving to death. The army reserve checks were helpful to me, and Rich was still kind of picking cotton here and there and yonder. Bryant was doing a little bit [also]." It was a smart investment.

The partners needed more brood fish in order to get more spawns for the growing business. They were able to buy some from Auburn University again, but an important source came to be the selection of fingerlings they had raised. They had already known about breeding practices commonly used to improve on characteristics in cattle and poultry. Beginning with the second year's crop, they began to separate out fish for the following year's spawn. "Every time we were grading fingerlings, when we'd see one that was twice as big, that went into the pond where we were raising replacement broodstock," according to Stephens. They did not know exactly what they were doing, but at the time, it just seemed like a good idea. It paid off: "The whole time the STRAL Company existed, we were going through the fingerlings and taking out those that were the largest and best-looking. Our fingerlings were head and shoulders [above others] in maneuverability, feed conversion, and dressing weights."[2]

Each year, the hatchery raised fingerling crops that were larger than previous ones. The numbers of STRAL spawn and fingerlings increased considerably when the men started spawning the fish in containers in the ponds. Spawning in the aquaria had clear disadvantages. Only so many aquaria could fit in the barn, and the whole procedure required a great deal of the men's time and energy. Water had to be pumped through the aquaria constantly, and spawning season was prime time for thunderstorms and electrical outages that could put the entire process at risk. It was likely in 1964 that they obtained some snow fencing to build containers for the ponds, according to Stephens. "We took the snow fence and drove each little stake into the mud, and we'd have a pen. We started off with maybe six pens. . . . We'd put a male in each pen, and then we'd put a female in the pen with the male. We'd have a can there for the female and the male to spawn in. And we still would use the aquarium, but it was much easier if we could spawn them in the pens, because you'd have to worry about the water going off and everything in the aquarium."

Not far from Newbern lived a man by the name of "Fuzzy" York. Fuzzy had a lot of old ten-gallon milk cans not being used, and he offered his friends at STRAL as many as they wanted. Not much later, on a trip to see the Auburn University football team in New Orleans one year, Stephens and True found the very same type of milk cans being sold as antiques for about $150 each. What a surprise it was to realize that they had cans worth that much lying around on the mucky bottoms of their fishponds!

Alabama's catfish production started largely as a family-based industry. Running the STRAL hatchery was a great deal of work, and family members often helped. While the Stephens and True children were too young to help much at the time, their mothers, Minnie Steele Stephens and Stella True, were actively involved. When they were not busy caring for the children or taking care of household business, they answered endless phone inquiries and delivered messages. When the men had to be away during spawning season, Stella checked on the eggs and the fingerlings, making sure the water didn't overflow and the tiny fish were not lost. Occasionally, she helped count fingerlings for delivery.

The job often involved heavy labor as well as long hours, and for that the partners sometimes needed help. All three men were used to working hard, and none of them shied away from physical labor, but when time came to seine and load large quantities of fish, they hired a young man who lived nearby. For the most part, though, they did whatever was needed for the hatchery themselves, working often from dawn to dusk or even into the night. By the second year, catfish-related work often took almost half their time. The responsibilities of raising and selling fingerlings started to impinge on time needed for the partners' "real" jobs.

The men concluded that something had to give. By 1964 True decided to put himself full-time into the fingerling operation. He sold his dairies and leased out his farm. The change was not a major one for him, since the hatchery was located on his property, and he continued to be the primary caretaker of the operation. Allen did the same with his farm. Stephens still sold feeds but found himself putting ever more time into the sales and delivery end of the fingerling business.

The venture had succeeded at establishing a reasonably solid economic footing. The men were splitting what they made among themselves, but the hatchery could not provide a living for all three of them. When the possibility came up of working as an engineer with Alabama Power Company, Allen chose security, a steady income, and a job for which he had originally been trained. "Probably didn't have to kill himself by working eighteen to twenty hours a day like we had been doing," comments Stephens. True and Stephens bought Allen's share of the hatchery and managed the entire operation themselves. The size of the workload varied sporadically. While they were not able to provide regular work for anyone else, they found extra help when they needed it. Even though it meant very long days, they continued to do all the deliveries themselves.

In those early years, catfish production was still such a novelty it could hardly have seemed like an industry. Just as spawning and production were a constant learning experience, so was running the business end of STRAL. The partners prided themselves on having good relationships with their customers and on dealing honestly and fairly, but they did run into problems on occasion.

One such problem occurred when customers would change their minds about orders. "About the second or third year, we had a terrific amount of people wanting information and giving us orders," Stephens remembers. "We'd call to deliver, or get there, and they'd decide they didn't want them." The partners came up with a plan: "We sent out a mailing to everybody about booking fingerlings. They had to pay cash in advance. One half a cent for every fish they wanted. If they wanted a thousand fish, they had to send in a deposit. That used to worry my daddy to death. He'd say, 'What if you can't fill it?' Well, then we'd give him his money back. But we were always able to fill the orders. And then when we had a cancellation, we kept the money."

Later, the deposit for each fingerling was raised to a full penny. "It was the best thing we did," remembers Stephens. In 1963 or '64, "one fellow up in Illinois gave us a nice order. He gave us a deposit of six hundred dollars. He ordered a bunch. That six hundred dollars was right at a critical moment in the operation, because we needed the money desperately. It kind of got us through." The men felt good about their accomplishments. Through trial and error, with determination, hard work, and the help of friends, they had succeeded in establishing a business in which they got paid on delivery and seldom, if ever, received complaints.

There was no way to be prepared for every problem the operation was going to encounter. There would still be times when the men needed something before they even knew exactly what they needed. They knew of no commercial catfish fingerling producers to consult. They had no books describing commercial catfish production techniques and no catalogs offering equipment and supplies for their purposes. They continued to rely on their own wits and on the knowledge and experience of researchers and fisheries experts in Marion and Auburn.

~

There were very few repeat customers during STRAL's first six years. Once a farmer's ponds had been stocked, he would not order more until the fish he raised were grown and had been fished out, and that took a while. As time went on the size of the orders coming in became much larger, however. People began asking for deliveries of as many as fifty thousand fingerlings at one time.

The hatchery had to expand if it were to accommodate the growing demand. The owners went back to the bank in Selma that had issued their first loan and borrowed money to pay for more ponds. They needed ponds for more fingerlings, and they needed an additional reservoir to hold water from a second well they had drilled. By 1964 the men had built additional ponds on True's property and leased between eight and ten ponds from Cooper Henry, a farmer with land a few miles away. The additional water helped increase their production to two to three times what it had been before. They paid the loan off within less than a year, and by 1966, they were in a much better position to fill the orders that came in. From that point on, they began making a profit.

Notes

1. True, interview.

2. By "feed conversion," Stephens was referring to how much weight the fish gained with the feed they were given. (See chap. 11, n. 3, for a slightly more technical explanation.) By "dressing weights," he meant the amount of the fish's weight that could be sold for food after processing (i.e., skinning, boning, etc.).

4

Early Producers

> In fewer than 30 years, seafood may become out of reach for millions of consumers. The National Marine Fisheries Service reports that more than 40 percent of fish species in U.S. waters are known to be overfished or seriously depleted. And the United Nations' Food and Agriculture Organization reports that the catch of some species has fallen by as much as 80 percent over the past two decades. . . . And while seafood supplies are on a serious decline, seafood popularity and consumption, as well as the overall population, is on the rise.
>
> —The Catfish Institute

When Mary Ezell's husband, C. A. Ezell, was a commercial fisherman in Alabama during the second quarter of the twentieth century, the rivers still contained plenty of fish. He and his crew of three or four men would go out to fish the Tombigbee River, the Alabama River, and the Tennessee River. He sold the fish they caught in what was called a "fish house" he had built on the hilly bank of the Tombigbee, just downhill from their house. The business grew. "It would seem funny to anybody today," Mary recalls. "The house was kind of on a hill then, but the bank is all gone, mostly, now. We had boxes that we sold fish out of—a retail fish market right there on the river bank. Then we had a big wholesale business. He sold to other restaurants before we got into the restaurant business in 1954."

Times began to change, however. "It must have been sixty-two, -three, or -four. Fish was getting scarce," recalls Ezell. "We didn't have enough fish to supply everything. Before we started using pond fish, he kept saying that the supply was running out." According to Tennessee Valley Authority estimates, the harvest of river catfish declined from more than 5.2 million pounds in 1963 to 2.8 million pounds in 1967.[1]

Is it any wonder, then, that as river catfish became more difficult to find, farmers paid special attention when the STRAL operators talked to them about producing catfish? By the mid-1900s many farmers, because of insects, weather, or economic fluctuations, weren't able to do much more than make ends meet. It did not take much to bend their ears when Check

Stephens and Richard True talked about stocking their ponds with fingerlings. "You can make some good money out of this," they told people. Some of these early ventures were later among the largest, most successful catfish farms in Alabama.

Not everyone took on great risk investing in this new kind of farming, however. Early on, farmers began growing catfish only as a part-time operation. People still had cattle, dairy cows, or row crops, and fish were to be only a sideline. Most farmers started on a small scale, typically, with their existing ponds. Everyone had thought catfish farming would be just a small part of the local economy. Their intent was only to give it a try, have some fish for their families, and perhaps earn a little money on the side by making their ponds accessible to the public for a fee. They hoped it would fit in with their other farm activities. At least, according to Elvin Wright, "if it didn't work, people could say, 'Well, we won't lose our pants.'" In Browns, Alabama, for instance, local residents and visitors could place one dollar in a mailbox on the road in front of the home of William W. Weissinger, a farmer and veterinarian. In exchange, they could fish in any of his four ponds all day long and keep everything they caught. Weissinger had stocked the ponds with bass and bream. In 1969 he drained the ponds and stocked them with catfish.

Most Alabamians who began producing catfish in the early 1960s were in the western part of the state, south of Greensboro, where the soils are rich in clay and hold water well. Safford, Alabama, is one such place, where brothers Harry and Neal Miller kept and sold bees, taking off the honey, and raised cattle and hay for well over thirty-five years. In 1964 their uncle O. W. Caine put seven thousand STRAL fingerlings into a three-acre pond. They watched him ride around the pond in a small boat, scooping feed from a bucket and throwing it into the water. The feed sank to the bottom. They had no idea what would come of the experiment. In 1965 their uncle harvested seven thousand pounds of catfish and opened his pond to people for fishout, receiving thirty-five cents for each fish that was caught.

The brothers became interested from the very first day they watched their uncle. They decided to try raising catfish for themselves in an old pond on their property that held bream and bass. The Millers stocked their first fingerlings in 1966. It was simple. All they did was feed the fish. The men who worked nearby on the railroad came and sat around the pond in the afternoons and fished. "It was mighty good," according to Harry, the older of the brothers. "Everybody would come out to visit who wanted to fish. We enjoyed it." The fish provided a source not only of recreation and sport but also social life and income for the Miller family.

Lester Taylor, and his son, Milton, who was in high school at the time, were among the only folks north of Tuscaloosa who raised catfish in those early years. The Taylors had a row crop farm in Town Creek. They decided to build a pond on their land, a place where family members could enjoy fishing. In 1963 Lester Taylor went to the SCS for help in laying out the pond. As was done in those days, a dam was constructed at one end of a hollow between some hills. Over the months, rainwater collected and filled the depression, and the Taylors were able to have STRAL fingerlings stocked. With the delivery, they received a pamphlet with guidelines on how to feed the fish as they grew.

The family enjoyed the pond and the fish later that year (see fig. 4). The pond had more fish in it than they could use, however. The logical conclusion seemed to be to open the pond to the community. They got word out through the small local country stores: "If you like to fish, we've got a fishing pond." The message spread. Community members, but also a few visitors to the area, came to fish, and paid for what they caught. According to Lester Taylor, they "had plenty of customers. They took up what fish we had, so it was a success."

When it got cool that fall—and Alabama winters are fairly cold in the north—the Taylors prepared to drain their pond to collect and sell any remaining fish. They placed large tubs under trees on a little hill beside the lake and filled them with ice. Milton Taylor recalls, "Everybody wanted to see what the pond looked like drained, with all those fish in it. There was a lot of interest. Several people came around, and they'd help tote the fish out. Some came and helped dress them. You've got a lot of interest in something new." The Taylors figured out how to skin the fish quickly with pliers and iced them down. They froze what was left after sales. The ease with which they later sold the fish surprised them. The next year, the Taylors restocked the lake and opened a fee-fishing operation. Their farm in Town Creek, Alabama, still produces catfish today, primarily for local customers.

Just thirty miles from Georgia, about twenty-five miles east-southeast of Troy, lies the small community of Clio. There, William Easterling became involved in the catfish industry in a very different way.

"Take a look at these channel catfish eggs," said Ellis Prather one spring day in 1963, lifting a jellylike mass out of the water for Easterling to see. A

Fig. 4. Lester Taylor with grandson at first family fishpond, summer 1978. Lester Taylor and his grandson Bart standing on the bank of the family's first pond. They are looking at a catfish they caught with a rod and reel. (Photo courtesy of Nellie Taylor Estate, Milton Taylor, Personal Representative.)

preacher and occasional lawyer, Easterling was visiting Auburn University's fisheries program and the research ponds north of the campus. Prather was describing to him the work he was doing with catfish. Easterling recognized the potential in what he saw and heard. "This is going to be a developing industry," he commented. "Someday, it'll be something real big."

Detecting a ripple of interest, Prather tossed out the line.

"Well," he broached, "why don't you become involved in it?"

Though not taking the conversation all that seriously, Easterling nevertheless took the bait. He responded with a casual "Sure." He added agreeably, "We've got some property with a lot of water that might be suitable to do something like that."

Prather pulled the hook a little tighter. "We could let you have some brood fish," he offered.

Easterling thought briefly. Certain circumstances at that time in his life might favor such an undertaking. Although the ministry had always taken priority with him, pastors at small churches needed a second job in order to be able to support themselves and their families financially. He was not really interested in continuing the little bit of legal work he was doing. In addition, he was looking for a way to arrange things so that his elderly father could stay at home rather than go to a nursing home, and in order to do that, he felt he would need to have people and activity around the house.

"Well, all right then," he said, "we'll take fifty pair."

After the visit to Auburn was over, Easterling drove south, home to Clio. He got busy with his ministry and other involvements and promptly forgot about the conversation. Six months or more went by.

One day, he received a phone call. Ellis Prather was on the other end.

"We have your brood fish ready for you," Prather said. "Can you come up tomorrow and get them?"

Easterling was surprised. He had forgotten all about the exchange with Prather. The call had caught him totally unprepared. He hedged a little.

"I don't have anything to haul them with, and I don't have a pond to put them in."

"Well, we can hold them for you a day or two," Prather offered helpfully.

What could he do? He had to accept. After he hung up the phone, questions raced around in his mind. He picked up the telephone and got in touch with a first cousin.

"Dan," he said, hoping this would be the solution. "I got myself in a real tight bind. I committed myself to get fifty pair of brood fish, and I don't even have a pond to put them in. You're always trying to help somebody. How about helping me?" he said.

"Sure," his cousin promised. "Go ahead and get 'em. I'll help you eat 'em," he joked. They laughed together, although there was a touch of nervousness in Easterling's laugh. Then Dan offered his cousin the use of a small pond on his land. "You can keep the fish there until you've got your own pond."

"Thank you, Dan," exclaimed Easterling gratefully. "I owe you. You just saved me!"

Easterling was a positive thinker. Spawning and raising catfish fingerlings looked like it was going to be an interesting experience, he thought, once he found himself headed in that direction. He viewed it as something like a hobby. Easterling himself would naturally continue to focus on his duties as a preacher.

But after Prather's phone call, Easterling's life began to change. Although he was unprepared at first, he wasted no time gathering information from every possible source on how to breed channel catfish and raise fingerlings effectively. With the help of the Alabama Department of Conservation and a local contractor, he began to design and build a small pond. He copied as much as possible of the design and techniques he had seen at Auburn. Quickly, the half-acre pond was built. Fifty pens were built and placed in the pond.

In nearby Ariton, where Easterling was pastor, a local teacher offered to have his students help Easterling build what he needed. "If we can help you in any way, let us know. It would be good for our boys," Mr. Steele said.

"Well," said the pastor-turning-hatchery-operator, "it would certainly be mutually beneficial for us all. It would give them the opportunity to see a developing industry in the pioneer stage."

Easterling worked out the design for the hatchery equipment he needed and gave it to Steele. The students not only built Easterling's first hatching troughs but also his first paddle wheel system. Next to the house in which his father lived was an old, two-story mill building that they set up as the hatchery. In it they put the paddle system and troughs in which they would hatch the fish. They changed the plumbing around so that the water ran through the troughs.

The next year, in 1964, Easterling hatched the first commercially spawned channel catfish fingerlings in East Alabama. The state Department of Conservation and AU fisheries researchers and extension personnel helped spread the word. The Easterling Fish Hatchery's first catfish fingerling crops went to people in surrounding areas who had fishout ponds and to people who wanted to grow the fish for their own use.

~

In the course of searching for ideas on managing a hatchery, Easterling had met Richard True and Check Stephens. They shared information, as did most of the people who were involved in spawning and raising catfish. Within two or three years after he had begun operations, Easterling had a customer who wanted to buy twenty thousand fingerlings. He didn't have enough to complete the order, and he asked the STRAL partners if they would help him by supplementing with theirs. At that time, twenty thousand fish was still such a big order that both Stephens and True came to deliver the fish with him. In the course of the delivery, Easterling remembers with a chuckle, they made a suggestion: "Why don't you let us hatch all the fish and you just sell them? You can be the distributor for us."

He knew that the offer was both in jest and serious. He was clearly heading toward being competitive in the catfish hatchery business. Easterling had invested too much time and money and had too much at stake to stop producing catfish fingerlings now, so he turned the offer down.

"You oughta put some catfish down there," Check Stephens was telling Frank Lee one day in the early 1960s. Lee was commissioner for the Alabama Department of Corrections, which managed the State Cattle Ranch, a minimum security correctional facility south of Greensboro. Land for the ranch had been purchased in the early 1940s, and the facility had been in existence since around 1948. During the 1950s and 1960s cattle was the ranch's main enterprise. The beef provided food for the state's prison system. The property had a few ponds with water from artesian wells, where the cattle drank and cooled off and inmates fished occasionally in their free time.

"Well, now, how do we get the money?" asked Lee.

Stephens shook his head and admitted he did not know. But raising catfish there would help provide food for the inmates.

"Well, maybe we can work something out," offered Lee.

By the mid-1960s the STRAL hatchery was delivering fingerlings to the Cattle Ranch ponds on consignment. The plan was for STRAL to buy the grown fish back and for the ranch to keep some in return for its investment. Before long, more ponds were built specifically for catfish, until there were thirty-eight in total. Ranch residents began spawning and raising fingerlings to use in the facility's own ponds and to sell to local buyers. For a short time, some of the fish were kept for a source of protein for the inmates, but that did not work out economically in comparison to what the fish would bring when sold on the open market. Catfish became the State Cattle Ranch's main income operation.

The STRAL hatchery operators worked together with AU researchers, providing a place where new findings could be tested in the field. Among the investigations done were studies of varieties of catfish other than the channel catfish. The researchers wanted to know whether any of them might have combinations of characteristics that would make them better suited for commercial production than the channel. In the early 1960s the white catfish, which is not actually white, was one object of attention. It seemed to show promise. "It could withstand oxygen depletion better than the channel," explains Check Stephens, "and it grew, we thought, a little bit quicker. It was a heartier fish, we thought at the time." The fish had a number of other features of special interest, also. "The white would spawn about two or three weeks earlier than the channel. We could sell the white fingerlings. We sold a goodly number of them. They did real good for fishing. The meat was good, the taste good. But the real big difference was in the dressing percentage, and that made it uneconomical."[2] For a number of reasons, the white catfish was deemed not promising enough. Channel catfish became the chosen species.

Joe Kyser Sr. was a neighbor and good friend of Richard True. Kyser operated a dairy and beef cattle farm, and the two men talked often about cattle farming. Ever since True had started the hatchery on his property, the topic of conversation turned increasingly often to catfish. Kyser watched his friend's business grow and heard about STRAL's predicaments, mishaps, and successes. His friend kept encouraging him to raise catfish himself, but the business was still relatively new and, he felt, very risky. Kyser was generally quite conservative in his decisions, but in 1965 he caught the bug and decided to go into catfish production in a big way. His intent was more ambitious than the other farmers who had begun to raise the fish. He wanted to start on a larger scale and sell the fish commercially.

First, however, Kyser had to obtain the money to do so. He approached the Federal Land Bank for a loan to buy property on which to build the ponds. He went to the Production Credit Association (PCA) for a loan with which to build and operate the farm. The PCA was an agricultural lending agency that had been established with federal seed money in the early 1930s. Although Kyser had previous ties with the PCA, he was met with great reluctance. Loan officers in those days found themselves in a predicament when asked for money to finance a catfish farm. "Catfish was very new," explains Bill Baker, a PCA loan officer himself. One of their concerns was that "you could not see all of the catfish once you put them in the water, as you could a pine tree or herd of cattle. You could physically make an inspection upon those. . . . The only time that you knew the fish were there was when you fed the fish, and that in itself was something new. No one had ever dreamed that you could feed fish and that they would come to the surface and eat. So we had to come to some kind of methodology as to how we would collateralize our operating lines for the production of catfish."

Part of deciding whether or not to lend money involved working out a way of securing the loan. No one had any idea what value might safely be assigned to the fish themselves, so other things had to be identified as collateral. Much the same as today, securing a loan with a lien on cattle or crops was safer than securing one with a lien on fish. "It's a risky business because your product is so fragile," says Charles Thomas of the Citizens Bank of Greensboro. "They can all die one night. They can die slowly from diseases that are hard to treat because you can't pen them up. We always take a lien on the fish, but we don't give it very much value, because they can just all go away. So we look past the fish to the man's real estate and his equipment, generally, to secure a fish loan."

Banks were naturally not eager to lose their customers' savings on failed ventures, and they were being asked to give loans on something about which they knew almost nothing. Five basic credit factors were considered at the time in giving loans, according to a 1970 article in the *Catfish Journal:* the man and his family, the man's financial position and progress, his repayment ability, the purpose of the loan, and the collateral. Not all these factors were objective ones. Loan officers tried to determine applicants' honesty and trustworthiness. For the earliest loans, they relied primarily on these subjective criteria to assess the risk. In small towns such as Greensboro, banks were able to do that fairly successfully, explains Thomas. "That's the advantage of being a banker in a small town bank, especially where you grew up," he says. "That had more to do with who got loans than any kind of formalized program. . . . Obviously, we'd still look at what they were proposing . . . see if we thought it was reasonable. If it was, then we'd make it. There's an old adage: 'You loan to the man, not on the land.' If you're dealing with an honest, hardworking kind of person, they're going to find some way to pay you back."

Joe Kyser mortgaged his entire dairy farm and beef cattle operation to establish the first large commercial catfish farm in Alabama in 1965. His youngest son, Bill, remembers how they harvested the first crop of fish, taking them "up the hill in wire baskets from the water and pouring 'em on the back of a trailer that we were pulling behind a farm tractor, drove 'em to the highway with the tractor, then loaded 'em into the tractor trailer truck on ice—just the live fish on ice—and sent them to the Chicago Fish Market."

In a small town like Greensboro, most people know, or at least know of, the other residents. Joe Glover owned and operated a local grocery and meat market on Main Street in downtown Greensboro and knew Richard True from his high school years. Check Stephens had been living in Greensboro for a number of years. Near the end of 1964 the three men began what was to be a fateful conversation about the potential of catfish farming in Alabama.

Joe Glover had come into the world in 1928 and grown up on the family farm near Greensboro. It was a small farm on which the family raised a variety of products—cows, hogs, cotton, corn, sugar cane for syrup. In 1955, after spending a few years in the army, Glover had the opportunity to buy the grocery store where he had worked evenings and Saturdays as a high school student. It was a family store, as the stores were at the time, with a meat market and a cooler and small freezer in the back. Joe Glover knew the meat business, and his customers knew they could get good meat from him.

An avid fisherman, Glover had always been an outdoors person. He fished every opportunity he found. "I would put out trot lines to catch catfish, and I would fish for bass or bream or anything in the creeks or streams that I could, because I enjoyed fishing," he recalls.

Glover also read a lot. He read magazines, and when he came across publications and information from Auburn University, he read those too. He had seen articles about catfish in farming magazines, and when he saw some AU publications about the promising future of farm-raised catfish, his interest was piqued. He read about the experimental work being done with catfish in ponds. He knew that the fish were available, because he knew Stephens and True had recently begun to hatch fingerlings.

Glover made a decision. There was a natural hollow—a small valley—on his homeplace farm. If it were cleared and dammed up, he could have a pond. When he asked the SCS engineer to survey the area, determine the actual acreage, and recommend how the dam should be built, the agency helped him just as it had others who wanted to start raising fish. Many trees had to be cut on the heavily wooded property, and their trunks and debris had to be removed. When finished, the pond turned out to be between an acre and an acre-and-a-half in area and very deep at the lowest point. In late 1965 or early 1966 Glover called the STRAL operators and told them the pond was ready for their fingerlings.

Glover hoped that he would be able to sell the fish and make a profit from the sales. More than the economic payback, though, he enjoyed the process of raising the fish. He usually went to the pond late in the afternoon, after he and his wife, Doris, who worked with him, had closed the store. Every day, he went to feed the fish and check the oxygen. He looked forward to going out to the pond with the bag of feed. Glover enjoyed watching them come to the surface for the pellets he threw out to them.

Glover's son, Joe Jr., or Joey, fed and watched with him. He would take Joey and sometimes Joey's friends to the pond and enlist their help to keep track of how the fish were growing. With cane poles, Joey and young Charles Thomas or other friends would throw in lines and pull up a fish when they felt a bite. Checking its size, they then walked home, carrying poles and fish to check the flavor for Mr. Glover by enjoying a meal of fried catfish. In later years Glover would invite friends and acquaintances, "Anytime you're out this way, catch a few and eat them and see if they're good to eat, if they're all right." They always were.

Notes

1. Cox, "Catfish Farming Industry of Southeastern United States," 39.
2. The dressing percentage is the percentage of the fish's weight that is marketable after processing.

5
What Now? Enter Processing

> There were too many fish. Everybody had a fishout pond. There was no public to go fish.
>
> —Richard True

"What am I going to do with all these fish?" farmers would ask Richard True and Check Stephens. "I've got fish in the ponds, and I don't know what to do with them all!"

The STRAL hatchery's business had been growing steadily, and the success brought with it unforeseen challenges. Catfish farmers were finding quickly that they had more fish than they knew what to do with. They turned to the fingerling growers for help.

"Well, fish 'em out," Stephens and True had suggested at first. That had seemed a logical thing to do. "Why not make your pond a fishout lake and advertise to folks that they can come and catch catfish to keep for so much per pound?" Many ran ads in local newspapers and opened their ponds to the public for a fee. That worked for a while, but after only a year or two, fishout catfish ponds had become so numerous that farmers could not sell their fish.

A few farmers made a small dent in their supply by selling some of their fish through local markets, but the sparse rural populations could not consume all the fish that were available. From his feed sales experience, Stephens recognized that the farmers did not know how to begin to reach larger populations: "Nobody had gone in and organized anything for the marketing of the fish. The manufacturing plant, when they build automobiles, they got a dealership that they can go to to sell them. And when Kellogg's makes cereal, they know where it's going to be sold. The farmer grows something, and a lot of times he doesn't know where to go sell it."

When farmers went to the fish markets, the markets offered them ten or fifteen cents a pound. That price came nowhere near even covering their costs. After all, fishermen were still pulling catfish from the rivers and selling them, without having the expenses of building ponds and paying for huge amounts of feed.

The hatchery operators began looking for a way to help the farmers market their fish. They talked with fisherman C. A. Ezell of Lavaca, who had bought STRAL fingerlings to raise and process for his restaurant in Choctaw County. There was no market for catfish there. Stephens traveled around and spoke with other potential buyers. "I'd go and check all the fish markets," he recalls. "I talked to various and sundry people. . . . They weren't going to buy any fish from us as long as they could get them out of the river for nothing. I went to Birmingham—the same thing. The same thing at the Florida Fish Co-Op, and Montgomery: 'We'll give you fifteen cents a pound for 'em.' Well, goodness alive, that didn't sound like too good a deal!"

The market was saturated. As quickly as it had bloomed, the new catfish farming industry now appeared in danger of collapse.

Stephens and True felt sure the fish could be sold for more than fifteen cents a pound. They continued to promote pond-building for raising the fish. In 1965 they even risked building a few ponds for customers with the idea that, after a year had passed, they would harvest the fish themselves. They were aware of the well-established poultry industry in Alabama and felt there was similar great potential for a catfish industry in the state, but they themselves were in no position to help all their customers sell all their fish. If only there were a facility that would buy the farmers' fish to process and sell, such a plant would be in a much better position to find ways and places to sell the fish. The men were convinced that a potentially solid market for catfish existed.

Among the many individuals who had built a pond and stocked STRAL fingerlings was Joe Glover. Glover was an accomplished butcher and knew the meat business well. A quiet and thoughtful man, he had been watching and reading about the fledgling catfish farming industry for some time.

Stephens and True began talking with Glover about the idea of starting a catfish processing business. Glover was very interested. The three agreed that such an industry would have great potential. Each of them would bring special experience and abilities to the undertaking. Richard True knew how to operate a hatchery and produce fingerlings. Check Stephens was an effective salesman. Glover knew about processing meats. He was adept at operating the equipment that would be needed, and the modest meat-processing facility in his store would be a good place to give it a try. Together, they had the combination of knowledge, experience, and ability to make such a project work.

The men needed a better sense of what they might be getting themselves into and what might be required for the undertaking. They investigated existing processing operations, looking especially for fish processing plants. They visited and spoke with people Stephens knew in Birmingham who had been buying ocean fish and dressing them in their plants.

Glover had read about an Arkansas baitfish farmer by the name of Edgar Farmer who had started a co-operative fish processing plant a year or so earlier. He had been processing buffalo fish grown in rice paddies. The three men traveled to Dumas, Arkansas, to observe how the plant there was operating. Although the Dumas plant was designed and used primarily for buffalo fish, the men did see a few catfish being hand-processed.

The three had many discussions and questions about the potential problems associated with starting such an enterprise. What size ponds should be built? One acre? Larger? "Not knowing, you don't want to venture too far out on the limb, you know," says Elvin Wright of the Soil Conservation Service, who helped build many ponds in the area. He remembers other questions the men were asking: "If a plant was going in, how many fish could they sell? Nobody knew. And where are you going to sell this product? Who's going to buy? And are the fish going to be good every year? Are you going to be able to keep a good product out there? That we didn't know." Finally, one day in 1966 the matter was resolved. "Let's just go ahead and do it," Glover proposed. "Let's get into the processing business."

So they did. They organized the STRAL Processing Company with Edgar P. Russell, a Selma lawyer with ties to Stephens' wife's family. The business was organized separately from STRAL Company. The plan was to process and market catfish grown by Alabama farmers as well as to promote the growing of catfish that could then be marketed through their plant. The first fish they planned to process would be the first ones Joe Glover had raised in his pond. "We didn't have any money much in it," Stephens recalls. "Each one of us took 15 percent of the stock in it. Joe put in the labor to run it. We used his store—that was his part of it. Richard, Edgar, and I put the money in it, because we didn't have any physical assets in it." And so they made their preparations for processing in the back room of Glover's Grocery.

Greensboro, Alabama, early Monday morning, November 14, 1966: The morning sun, still below the treetops, threw long shadows across Joe Glover's small pond north of town. Today, the fish were going to be harvested by STRAL Processing Company. It was to be the first harvest of the pond and the first processing by the new partners. Since early that year, with help from his teenage son, Joey, Glover had fed and watched the three- to four-inch fingerlings grow, some to more than a foot in length. Curious onlookers were gathering among the trees alongside the bank, talking and laughing among themselves. There was a bright air of anticipation. Four days before, Glover had placed an ad in the *Greensboro Watchman,* the local newspaper, that read, "FRESH FISH SALE. Joe Glover will drain his Catfish Pond on Monday, November 14. Everyone is invited to buy fish at the pond at 50¢ lb." It seemed as though everybody in the community had come to see the fish harvested and to buy some.

The dam was forty feet high on the downhill side. A farm truck turned off the road and rolled slowly out along the top of the steep embankment stretching between the wooded slopes. The driver brought it to a stop at an agreed-upon spot. Two wooden tanks full of water sat on the bed of the truck, each fitted with an agitator on the top to oxygenate the water. Joe Glover had had the tanks built specifically for harvesting catfish.

A second truck rolled up to the bank. This one had tanks on the back too, but these tanks normally carried fingerlings for delivery. Check Stephens and Richard True climbed out of the cab and donned their rubber chest waders. With them was Dave Thornton, who often worked with them at the hatchery. Today, the tanks were going to carry grown fish.

Joe Glover had gotten out of his truck and was pulling on black rubber boots. With him were five or six men he had hired to help with the harvest. There had been no problem finding men willing to help. Everyone was curious to see what harvesting these fish would be like.

Two or three days before, Glover had opened the drain that was built into the levee. The pond was small in surface area, but it was twelve feet deep at its deepest point. It held a lot of water, and Glover knew that draining it would take a long time. Over the past couple of days and nights, water had been gushing out of the drainpipe on the outside of the dam, but now the flow had slowed to a relative trickle. The water in the pond was finally low enough for the harvest to begin.

While they had watched the water level go down, it had become apparent to the men that the harvest was not going to be simple. Tree stumps of all sizes stuck this way and that out of the pools that remained. The men started down to the water, sloshing their way through the thick, reddish mud of the exposed bank. Although they had discussed and planned to

Fig. 5. Pulling fish in with seine at first STRAL harvest, November 14, 1966. Pulling fish to the bank with a seine at Joe Glover's one-acre farm pond during the first STRAL harvest. *Left to right*: Joe Glover (head out of picture), Richard True, unknown laborer (back to camera), Check Stephens (wearing glasses), unknown laborer. (Photo by R. C. Nelson, USDA, Soil Conservation Service, now Natural Resource Conservation Service.)

some extent what they were going to do, they had no idea how things would go. First, they drew the seine through the open water to collect the fish into one area.

The crowd on the bank perked up. The remaining pools of water were broken by the dark gray backs of catfish, alarmed at the men invading their last refuges. Occasionally, one of the onlookers called out a friendly, joking comment to one of the men in the water.

"Look, mama! There's some!" a young child told his mother and pointed. "See 'em?"

In the few spots that were clear of stumps, some of the men dragged the seine with a great deal of effort through the water, pulling the ends together toward the bank to contain the trapped fish (see fig. 5). With dip nets, others scooped fish out of the water and waded with a net full of flopping fish over to the bank. Turning the fish out of the net, the men dropped them into metal washtubs. When a tub was full, two men grabbed the handles on either side and dragged it up the steep earthen dam to the truck. Every few feet, they had to set the tub down, change their grip, pull their feet up out of knee-deep mud and haul the tub a little farther. At times, it seemed as though they slid down and into the mud as much as they went up. A small generator vibrated noisily on the back of the truck, powering the agitators in the tanks. The men lifted the washtub up onto the top of the truck's side. From there, others placed it onto the edge of one of the tanks and turned it over until all the fish had dropped into the tank. That done, the men trudged down through the mud, back to what was left of the pond to collect another tubload.

Before long, the harvesters realized they were not making significant progress. There just was not enough open water in which they could use the seine effectively. Too many tree stumps were in the way. Glover, True, and Stephens discussed the situation briefly and then spoke with the others. They began wading around in the shallow, roiling puddles that were left, bending over and lifting the fish carefully, one by one, out of the mud (see fig. 6). Slogging their way through the heavy mud was nearly impossible in places. Sometimes, they sank in up to their knees and had to pull their legs up and out of the muck one at a time to make their way to the boat and the washtubs waiting at the water's edge. Where the mud was exceptionally deep, stepping high did not get them out of its grip, and they could barely move. Occasionally, one of the men had to stop and help pull another man out. As fast as the boat was filled with fish, it was emptied (see fig. 7).

The sun climbed high in the sky. The crowd of people watching from the bank thinned somewhat. Some had already bought live fish that had been weighed right there beside the pond. Fish by fish, the tired harvesters went on filling the tubs and the boat that was now stuck in the muddy bottom. Stopping was out of the question. Tub by tub, they filled the tanks on Joe Glover's truck until they could hold no more. Once it was full of fish, the truck was started up and driven slowly off the dam to haul its load downtown to Glover's store. The STRAL fingerling truck was then backed into its place. The harvesting went on, with the men pushing their way through the muddy puddles, picking up catfish in the dip nets where they could and by hand everywhere else. The fingerling truck's tanks gradually filled up. At times, it seemed as though they would never get all the fish out of the pond. They took occasional breaks to gulp down a drink or a bite of food but soon went back to work.

Fig. 6. Picking fish up one by one at first STRAL harvest, November 14, 1966. Fish had to be gathered by hand from remaining puddles during the first STRAL harvest. Joe Glover, in the background, is picking up and tossing remaining fish into the boat. In the foreground laborers are preparing to carry the washtubs full of fish up the bank to the truck. (Photo by R. C. Nelson, USDA, Soil Conservation Service, now Natural Resource Conservation Service.)

Meantime, in Greensboro, Joe Glover stopped his truck in the alleyway that ran behind his grocery store. He went through the back door and walked across the room to the big walk-in meat cooler that stood just behind the rear wall of the market. The employees followed him inside, carrying tubs full of fish they had removed from the tanks. Glover opened the cooler's heavy, wooden door. He and two of the men began to pour and spread ice onto the floor of the cooler. On top of the bed of ice, they laid out a layer of catfish. Most of the fish were by now dead or near-dead and did not flop around the way they had when they were taken out of the pond. The men covered the fish with more ice. After placing a second layer of fish, they shoveled yet another layer of ice on top of that. They continued layering fish and ice until the truck was empty. Glover swung the heavy cooler door shut and latched it. One of the employees drove the empty

Fig. 7. A boat full of catfish at first STRAL harvest, November 14, 1966. This boatload of catfish was collected by hand during the first STRAL harvest. *Left to right*: Joe Glover, unknown laborer, Richard True, unknown laborer, and Check Stephens. Note the stumps, trunks, and branches remaining on the pond bottom, which made seining difficult. (Photo by R. C. Nelson, USDA, Soil Conservation Service, now Natural Resource Conservation Service.)

truck back to the pond, and Glover waited for Richard True to bring the fingerling truck. When he arrived, the second load of fish was taken off and iced down as the first had been, until the cooler was full.

The harvest continued until the sky began to darken. The men took approximately twenty-five hundred catfish out of Joe Glover's small pond that day. The STRAL Processing Company's first harvest amounted to about three thousand pounds. It was the most catfish anyone there had ever seen at one time.

On November 15, 1966, the first catfish were processed commercially for sale in Alabama. Ten or twelve people were at Glover's store by 7 A.M., ready to work—Glover, True, Stephens, and the men and women whom they had hired to help with this very first load.

Glover had built a tall, wooden A-frame and driven long nails through the board that spanned the length at the top. It looked a little like a tall sawhorse standing there in the back room of the store. As a boy, he had skinned catfish on a nail hammered into a tree, and the A-frame with nails along the top would serve the same purpose for the group. The processing began, with hand tools such as mallets, knives, pliers, and with the band saw in the meat market. "We hung the catfish on the nail," Glover remembers, "and then cut around the gills with a knife. Then we took a pair of pliers and pulled the skin off, and then sawed the head off with a band saw and eviscerated 'em, washed 'em, and re-iced 'em. We processed them just like we would when we would go fishing."

At any one time, between six and ten people stood around the A-frame, skinning catfish that ranged from ten inches to as much as two feet in length. Glover had taken the first few and shown the others how to do it. Then he took the skinned fish and cut the heads off with the bandsaw. He cut the fish from end to end with a long boning knife. From there, the fish were eviscerated by employees, who used their bare hands and fingernails to scrape the insides clean. They then washed the fish thoroughly under a faucet over a big, square iron sink in the back of the store. The cleaned fish were taken out to the scale that sat in the meat market, where another employee weighed out five pounds at a time. The fish were packed five pounds to a plastic bag and replaced on ice in the cooler.

The process went slowly at first. But gradually the group fell into the groove of a system; they all became more proficient at their jobs; and the pace picked up. Before long, the group was skinning and cleaning, on average, two fish a minute. From 7 A.M. until 5 P.M., for three full days, the processors worked in this way. Throughout the days, close to a hundred curious friends and community members came to see for themselves how the venture was going. It was the first commercial processing of any fish in Greensboro and the first commercial processing of catfish in the state.

A cardboard sign was placed in the window of the grocery store as soon as the fish had been packaged. Large red letters read: "Fresh Farm-Raised Catfish for Sale." Soon, almost everyone in the area had heard about the fish. Many came to buy them. The reaction was encouraging: "Mr. Glover, those catfish you sold us were absolutely delicious! I fried some up last night for supper, and the family just couldn't get enough of 'em!" a lady said from the other side of the case of meats.

Glover wiped his hands on his white butcher's apron.

"Thank you, ma'am. Glad you liked 'em," he said with his modest smile.

"We've got more in back," he offered, gesturing over his shoulder toward the storeroom.

A man in overalls came down the aisle. "Well, I'll have some more!" he grinned. "You know, Mr. Glover, those fish tasted a far sight better'n anything I ever caught in the river!"

The STRAL partners had hoped that they would be able to sell the fish as fast as they processed them, but that did not happen. Five, six, seven days passed, and catfish still sat in the cooler. Some had to be hard frozen, because they were in danger of spoiling. Almost a month passed before all the fish were sold.

For about six months after that first experiment, the STRAL partners harvested fish from the ponds of only one or two other farmers. These ponds were larger and yielded more fish than they had taken out of Glover's pond. They sold the fish through the meat market, but it was clear that processing them by hand would not get them very far. There had to be a better way to do it. According to True, "We could never figure out how we were going to make it very big, because it was impossible to hand skin them. Unless we could find some mechanical way of dressing catfish, we didn't feel we could get very big."

In his work promoting and delivering STRAL fingerlings, Check Stephens had developed a friendship with an elderly man in Cordele, Georgia, who was in the meat processing business. In the days before the deep freeze had been invented, J. C. "Jake" Ledbetter had provided butchering and packaging services for people, renting them cold storage lockers in which to keep the meat they had purchased. Ledbetter was also very interested in the growing catfish industry. Since 1962 or 1963 he had been ordering fingerlings every two or three weeks from Stephens and selling them to farmers in Georgia. He was in constant touch with the hatchery operators.

One day in 1967, during a fingerling delivery to Cordele, Ledbetter mentioned to Stephens and True that he had a new machine, a model 66A skinning machine from Townsend Engineering Company that he had not yet used (see fig. 8). The machine was designed to skin beef livers and take the membrane off of beef tenderloins. Ledbetter asked the men how they were processing the catfish they harvested. When he heard they were skinning each one by hand with pliers, he told them he thought the machine might take the skin off for them.[1]

"I haven't paid for it yet," he told them. "I don't think it's going to be

Fig. 8. Townsend skinning machine model 66A.
The Townsend Model 66A skinning machine was first used on catfish in 1967 by STRAL Processing Company in Greensboro. This machine, which was originally designed to remove beef liver and tenderloin membranes, and later models modified for use on catfish were essential factors in making the processing of the domestically raised fish a practical and profitable industry. (Used by permission of Townsend Engineering Company.)

profitable for me to use. But I believe it'd skin a catfish. I really think it might work. Why don't you give it a try?"

"Well, that sounds like a mighty interesting possibility," Stephens responded, "but we haven't got the money to pay for the machine, and I can't come back over here with a fish just to try it out."

"Let me talk to Townsend," Ledbetter suggested. "I'll talk to 'em to see if they won't let you take it over there and see what you can do with it."

On the morning of his next trip to Cordele, Stephens got up as usual before daylight so that he would be able to make it back the same day. One of the tanks had been taken off the fingerling truck so that the machine could fit there if Ledbetter wanted to send it back with them that day. In Cordele, after the fingerlings had been unloaded into the large vats where they were kept, Ledbetter, Stephens, and one of Ledbetter's employees loaded the machine onto the back of the truck. The STRAL truck carried the skinning machine back to Glover's Grocery.

> The bodies of catfish do not have scales, but they have an outer protection consisting wholly of a thick, slippery skin. Spines that are sharply pointed and covered with a secreted "poison," protrude from the side fins of their bodies near the gills, and also from their backs, are a source of pronounced discomfort to anyone coming in contact with them.
>
> —*American Fishes and U.S. Trout News*

The first attempt at using the machine to skin catfish was a major event in downtown Greensboro. Stephens and others had sent word out that they were going to process catfish that day with a special machine, and a large crowd had gathered to watch. Tom Hooker, the representative from Townsend Engineering for the southeastern United States, and employees of Alabama Power Company were present. C. A. Ezell had brought some people with him from his catfish restaurant and fish farm in Lavaca, and others were there as well. People had come from miles away to see the new skinning machine do its work.

Joe Glover and his partners were prepared for the occasion. The skinner had been standing on the back of the truck in the alley behind the store. Donald Berry, a local employee of Alabama Power Company, had come to help hook the machine up, because it required a special, three-phase current, so it was wired-up inside the local utility company office, the only source of the required high voltage needed. A large tub of fish from a local farmer's pond sat next to the skinner.

The machine was designed to work with a roller that grabbed the liver or bacon rind and pulled it against a blade that scraped the skin or membrane off. Joe Glover took a catfish and tried sending it through the machine. It mangled both the skin and the fish. He tried again, a different way. Again, it did not work.

Testing new applications was not a new experience for Tom Hooker. Typical of the equipment development process, Townsend engineers took time to evaluate possible new uses and designs or modifications of the company's equipment, and Hooker's instructions on this occasion were to find out how to skin catfish. He offered to try. Before today, he had never seen a domestically raised catfish, much less tried to skin one with the 66A, but he positioned himself in front of the machine. According to Stephens, this was what happened next: Tom "put the fish on, took his hand to push it. . . . He put pressure down on the fish to make the skin come off." Then he guided the fish across the blade. That was when the fish, which still had some life in it, jumped. Suddenly, Check Stephens recalls,

> his hand slipped, and the blade took a chunk out of his hand about the size of a silver dollar.
>
> I grabbed him, looked at it, saw the blood. "Are you all right, Tom?"
>
> He said, "Yeah, yeah, I'm all right. It doesn't hurt." He looked down at it. He keeled over. So we picked him up. Wallace McAdorey's office, Dr. McAdorey, was right out the back alley. I said I was going to take him up there. They called Doc and told him I was on the way. So I grabbed him and held his hand and took him to Dr. McAdorey's office . . . went right in, because they knew we were coming.
>
> Doc sat him down there. He looked at me . . . said, "Where's the skin?"
>
> "I reckon it's in the skinner," I said.
>
> "Call down there and get it!" he said.
>
> One of the nurses called down there, and Richard and Joe looked through the fish skin to get his skin. They got his skin and brought it up there. Wallace grafted that back on him, and you know, it took!

There were no more attempts at skinning catfish that day. Everyone left the store in the accident's chaotic wake, shaking their heads in dismay over Hooker's injury and agreeing the machine seemed like it was just not going to work.

~

The STRAL partners, however, did not give up. They returned to the machine and tried repeatedly over the next several weeks to try to skin fish with it. They pushed fish across the rollers every which way, but the skin did not come off properly, no matter how often or how many different ways they tried. But they could not quite accept the fact that the machine would never work. Much as they regretted Tom Hooker's injury, the machine had, after all, taken his skin off remarkably cleanly—so cleanly, in fact, that although it had taken about 125 stitches inside and out, Dr. McAdorey had been able to sew it back so well that the graft hardly showed at all. Surely it had the potential to do the same with the fish.

Stephens said determinedly, "This dadgum thing has got to work some kind of way!"

True mulled it over. "You know, I believe it just might still work. It just might." he agreed thoughtfully.

When he had time, Glover kept playing around with the machine in the back room. Nothing seemed to help. Suddenly, one day, an idea came to him.

"Maybe the blade's not sharp enough!"

He took the blade out of the machine and asked one of his employees to take the blade outside and sharpen it. The man took it carefully and sat down with it on the sidewalk in front of the store. Although he had never sharpened this kind of blade before, he had sharpened knives on the concrete curb in the past. In the manner he used with knives, he carefully and repeatedly swiped the edge of the blade sideways against the concrete. Then he brought it back inside, washed it, and helped reinstall it into the machine. Lo and behold, the machine took the skin off quite neatly! After that, the men began to process catfish "left and right."

~

There were additional difficulties with using the Townsend 66A on catfish. The teeth on the rollers caught the skin and pulled it against the blade, as they were intended to do. But the skin was very difficult to get off from around the bony parts of the fish. The men had to use specially designed fish skin pliers on every single fish to pull it off, and so a great deal of work still had to be done by hand.

Another problem with the machine was that the fish skins would wrap up on the roller and jam the entire works. One day, while True stood washing the machine down with a hose, he noticed that the water prevented the skins and other debris from sticking to the blade. The three partners discussed the problem and came up with a simple yet ingenious idea: True took a piece of pipe and, after capping one end, drilled tiny holes along the remainder of it. He attached it with wires to the back of the machine, near the roller. When they attached a hose to the open end of the pipe and forced water through it, the water sprayed out over the roller. The skins slipped right off and fell onto the floor. A large bucket placed under the back of the machine would take care of those. The STRAL partners had found a way to make the skinning machine work on the fish. Soon afterward, they sent Townsend Engineering Company a check for twenty-five hundred dollars.

While STRAL was operating out of the back of Glover's Grocery, people from all over came to observe and talk about going into the catfish processing business. However, other processors had difficulty getting the 66A to work properly, and the STRAL partners helped get them started. "Several plants started up in Mississippi within a year or two," True remembers. "I sent people over there to show them how to operate the machine."

Through testing experiences such as the one in Greensboro, Townsend developed the Model 600 skinner to overcome the problems encountered with the 66A. One of the early redesigned models incorporated the water spray system that Richard True had originally jerry-rigged at STRAL to keep the skins from jamming the rollers. In addition, instead of teeth, it added among other things a continuous tooth roll, a stripper roll to keep the skin from wrapping around the tooth roll, a disposable razor-type blade rather than the resharpenable one, and optional air and water cleaning attachments. The company also furnished gloves for the machine operators to avoid injury. The machine gave a critically important boost to the entire industry. If there had not been some way to skin the fish, the mass production possible today could not have been achieved.

Notes

1. Much of the information about the Townsend Model 66A skinning machine was obtained from Tom Hooker in e-mail correspondence.

6

A Plant for STRAL

> Before 1968 the commercial processing of farm-raised catfish was almost nonexistent.
>
> —Gary C. Cox

Word about the processing operation spread quickly around the Greensboro community. People continued buying five-pound bags of catfish from Glover's Grocery. The STRAL partners actively encouraged farmers to buy fingerlings and raise the fish. Before long, many pond owners came to STRAL's door. Would STRAL please buy their fish from them?

The partners' space was limited. The walk-in cooler could hold only so many fish at one time, and the small back room, which still had to service the market's other needs, was cramped quarters for the group needed to process the fish. Joe Glover was already building a second, much larger pond, and other area farmers were building fishponds as well. What had started as a backroom catfish processing operation in November 1966 was promising within months to burst at the seams.

The next step was obvious. The men needed to build a plant specifically for processing catfish. Joe Glover's brother, Hoyt, owned some land on Centreville Road just inside the northern city limits. He was willing to sell it to the men, but they would have to raise money to pay for the land and to construct a building. They accomplished this by selling stock in the company. "I figured since we had a corporation we'd get the stock split and sell some," recalls Check Stephens. "We could have thirty stockholders total. So we split the stock and sold $30,000 worth. Some people put $1,000 in, some put $2,000, some $5,000, and we got to $30,000." People as far away as Chattanooga and Nashville, Tennessee, bought stock, but most of the investors in STRAL's future lived around Greensboro. The STRAL partners bought the lot and drew up plans for a small processing facility.

~

On August 22, 1968, page eight of the *Greensboro Watchman* sported the headline "Rotarians Hear about New STRAL Processing Plant." The local business community wanted to know more about the new enterprise and had invited Check Stephens to speak to the Greensboro Rotary Club. The article announced that the STRAL partners hoped to have the new plant completed and operating within one week. It was to process and store a hundred thousand pounds of catfish, in contrast to the three-thousand-pound capacity of the old site. "Most catfish come from Hale County and surrounding counties," reported the article, "though some come [from] as far as 100 miles away. Fish are brought into the plant alive and are killed at the time of processing and freezing. Mr. Stephens pointed out that this gives the user the freshest product possible. The product now has a registered trademark, 'Colonel STRAL,' and is sold throughout Alabama and to a limited degree in other states in the South."

While the business community hailed the establishment of the larger-capacity plant with enthusiasm, the reaction of the community was mixed. Some people viewed it as a boon, but others responded negatively. "There were a few people that would as soon nothing new came in," Stephens recalls. "People don't like change. You get used to something, you don't want to see change. Some of them would come up and say, 'Why wouldn't you want to put it someplace else?' "

There may have been some concern that the new plant would be competition for available workers. "Apparently, when something like this gets started, [there's] a lot of jealousy 'round about." Stephens explains. "You've got other people using help, and we were going to be employing lots of them." People had some concern also about the plant's nuisance effect on local air quality and sanitation. They "thought it would be a smell, from the standpoint of sanitation, flies, and all."

In general, however, the community supported the new business. Local government agencies provided services to help get it started. Municipal services were available in some ways that are no longer the case today. "They ran extra water out there," explains Stephens, "water lines and sewer lines to get us going. Back then, the county had road equipment. They had bulldozers, machines. They would go out to fix a lot. We didn't have to pay for it."

~

The new plant was far better equipped than the back room of Glover's Grocery. The new processors had to have everything custom built, because no equipment existed in 1968 that was geared specifically for the processing of catfish.

Much of the equipment used by today's harvesting crews is basically the same as it was then. The truck that carried harvested fish resembled, on a smaller scale, the type used today. Glover had a two-and-a-half-ton truck on which he had attached custom-made tanks. After seeing how exhausting and time-consuming it had been to carry fish up the pond banks, he installed a pulpwood loader—a cranelike contraption—onto the truck in 1967. According to Stephens, "This thing pulls up, with a cable on it that runs with a motor. You could let the cable out and draw the basket [of fish] in and load them in the tank."

The idea for the winch came to them through Milton Bowman, the Ralph Wells Company nutrition researcher. When he came to Greensboro from Illinois to help harvest fish to take back for his research, he set up a mechanism to bring the fish up to his pickup truck from the pond. It was rough, but it stuck in the partners' minds. "He jacked up the truck," says Stephens, "so that one wheel would turn, and took a spit, took the lugs off, and hooked it onto the wheel. Then he had a dolly run down to where the seines were. They'd load the fish up, crank the truck up. The truck would bring this load of fish up, and that's where we got the idea of winching them up."

One of the foremost purposes in building the processing plant had been to accommodate larger loads of catfish. The partners needed, therefore, a way to keep large quantities of fish alive until they could be processed. Always on the lookout for ideas, Glover recognized an opportunity when he heard of an old silo for sale. He bought the silo, which had been built in sections, each one about fifteen feet in diameter and four feet high. The sections seemed perfectly suited to be holding tanks. Glover's son, Joe Jr., remembers helping his father and other men place the sections on the ground next to the new building. They poured concrete into the bottom of each silo ring and created an impermeable floor that would hold water. With aerators, they now had tanks in which to keep a large number of fish alive.

Getting the fish *out* of the holding vats was a challenge, however. The only way they could get the fish into the plant was by draining the vats and grabbing and shoveling the fish out with pitchforks. Clearly, that was not a very good way to handle the fish, because it did far too much damage to them and was slow and awkward for the men. Later, the partners constructed tanks that were rectangular rather than round and used "crowder bars" to force the fish to the end nearest the building.

The inside of the plant was designed for basically the same procedures carried out in the back of the store. The only major addition was conveyor belts. Before, they had filled tubs with the fish after each step of the process and carried each full, heavy tub to the next station. Moving the fish along on the belts was a significant improvement. Other than the belts and the skinning machine, however, everything was still done by hand.

Even before they had conceived the idea of building a processing plant, the three partners had had to deal with important marketing questions: How should they try to sell their product, and how should it be processed to sell? The first fish processed in the store had not sold as quickly as the men had hoped. The partners looked to the highly successful poultry industry for ideas, but poultry processors sold their product fresh, and the men had learned early that there was simply not enough of a market for fresh catfish to do that. The only practical thing to do was to freeze it.

The frozen product's shelf life was anywhere from six months to as long as a year, and the frozen fish still tasted good when cooked. The question was, Which method of freezing should the processors use? Conventional freezing was the simplest and most affordable method. They obtained plastic bags in which fish could be frozen individually. The bags boasted a picture of their trademark, a whiskered catfish, and read: "Grain Fed, Straight from the Plantation. Colonel STRAL, the Aristocrat of Fresh Water Catfish." The washed fish were placed into the bags, which were sealed and put on trays to go into one of the new freezers. The moisture in the bag put a protective glaze on the fish. Thus STRAL obtained shelf life for its products. Now the plant would not lose fish if a buyer were to change or cancel an order.

Some product changes were made in order to make the processed fish more saleable. "In the early days," remembers Glover, "we thought all we needed to do was clean the catfish and sell it whole." The three partners actually knew little about marketing the fish when they first started out. They kept their eyes and ears open to what people were saying, however, and tried to respond. When Stephens returned from talking to potential customers such as cafeterias, restaurants, and schools, the partners realized that these institutional markets wanted something boneless. It stood to reason, therefore, that STRAL should offer a boneless catfish filet.

Some of the catfish coming into the plant from farmers' ponds were quite large and therefore difficult to sell whole. Glover developed a way to dress them that made them more saleable, obtaining a boneless filet in the process. "We called them 'splits,'" remembers Stephens, "We started off with Joe taking and splitting the big fish in half. One of the halves had the bone." The other half had been fileted off the bone. One thing led to another: "Joe was fileting the fish, and we needed something that people could fry them with—a breaded fish. It was economically feasible for us to bread the fish, because you could get more for the end product. So you were getting a better profit out of it. Plus you would be able to get a bigger market. That was the reason for breading." Doug Jeffords, who had made the breading mix for Kentucky Fried Chicken, was very interested in STRAL's operation and developed for the processors a flavorful seasoning mix for breading. The half of the split that had the bone was sold also.

The filets did not sell as the men had hoped, however. Consumers were not yet interested in the filets. It was apparently too early in their experience with any kind of processed catfish. So the splits, the filets, and the breading were discontinued.

Restaurants and other institutional food service customers wanted to be able to offer fish by size. The partners listened to potential customers and tried to respond with different products to build markets for the fish. The processors thought about how they might accomplish grading, and another step was added to the hand-processing line. "We did it by sight, to start," Glover explains, "with little scales, electronic scales, people putting the fish on them. They got fairly good at that and were able to sort 'em pretty fast, the little ones going in this tub, and [about that] long going in this tub, and so forth."

The fish processing industry was so new in Alabama and the nation as a whole that no health codes defined procedures the processors needed to follow. "There were no health standards for us to go by. The health department couldn't even inspect us because they didn't know what to look for," remembers Stephens. Nonetheless, constant attention was paid to keeping the facility clean and sanitary.

The waste of the processing business—the heads, viscera, and any other parts that are discarded—is called "offal." Offal was a major sanitation headache, because it attracted flies and because it began decomposing in a very short time, especially in the warm southern climate. STRAL plant managers mounted fans on the doors to create air currents and tried a variety of ways to block the flies' entry, but keeping them out was a constant challenge.

Plant waste had to be disposed of every day. For a while, True volunteered his own land as a disposal site. "I buried a lot of it back here on my farm," he says. "I came every day and buried it. I dug a hole and buried it deep. I had a backhoe, my company did, and we'd just dig a hole deep as it would dig it, cover it over, move over a few feet, and dig another hole."

A nearby rendering plant was processing wastes from a poultry plant. STRAL obtained permission to take its offal to the rendering plant and add it to the poultry by-products. The volume of the catfish offal was not so large that the plant could not handle it. That solution was just temporary, though. As the volume of the waste grew, the partners had to begin trucking it to the local landfill. Residents along the way tolerated the passing smell.

Getting the work of the plant done was not a problem in the early years. Plenty of people were interested in working there. At the helm, the partners' skills and personalities were well suited to the various jobs that needed to be done.

Richard True was a hands-on kind of person, "never one to sit behind a desk," in his words. As a manager of the plant, he "would always go in and show somebody how to dress a fish, go in the freezer and stack boxes, or suggest better ways of doing things." When a farmer's fish needed harvesting, he never minded grabbing a seine and helping the harvest crew bring in the fish. He often took the STRAL truck on deliveries. In addition to actively participating in the plant's operations, he continued to oversee the goings-on at the hatchery, where he had hired people to carry out the day-to-day routines.

Although Stephens helped with a variety of plant work, he spent less time physically there than True. "I watched. I knew everything. I've done everything in the catfish industry except cut the heads off," Stephens says. He spent most of his time doing what he knew best. "I'd been involved in contacting people. I had connections all over, and I'd just as well use them making the contacts to merchandise and sell the fish."

By the time the plant was up and running in 1968, Joe Glover had left the grocery and meat business he had been operating for about fifteen years. Greensboro had not been spared the political turmoil over civil rights that was taking place in Alabama during the 1960s, and a boycott of his store had had a significant effect on business. Faced with the prospect of growing demands on his time from the new processing plant, he decided to devote his energies entirely to catfish.

Particularly in the early years of the business, Glover worked on the line, skillfully cutting fish heads off with the bandsaw or fileting the fish himself. His ability to do a variety of tasks while working at the saw were impressive: "Joe could stand there. He could talk to you and cut those things and make that thing sing," says Stephens. Occasionally, Stephens would be too busy to meet with the many potential customers interested in talking catfish. If he could leave the plant then, Glover would go out to help with the marketing. He set up distributors or got orders from fish markets such as Williams' Seafood in Birmingham. Much of the time, however, he did the work for which he was most needed: overseeing the plant's operations. He was constantly expanding his knowledge about fish farming and processing, and applying that knowledge to the activities on his farm, with other farmers, or in the plant.

~

At first, the plant did not get as many fish to process as they had hoped they would. In the very beginning, the harvesters took fish only from local farmers. On occasion, they traveled to Selma, roughly fifty miles from the plant. Once or twice they went as far as the outskirts of Birmingham, but driving the truck and crew to distant locations to harvest ponds was not feasible, economically or otherwise. Because they had not yet begun to promote catfish farming in an organized fashion, they often operated the plant far below its capacity and at first could not provide employees with a steady work schedule.

Before long, serious problems surfaced with the irregular supply of fish. In order to provide extra incentive to farmers to produce, the partners had promised them STRAL would purchase their fish once they reached marketable size. "You put them in; we'll buy them," the partners had said. That was a way of getting their business started, according to True. But harvesting the fish exactly when the farmers thought they were ready and wanted them taken was not always possible. Sometimes, when farmers wanted their fish harvested, STRAL had no customers interested in buying them. Other times, customers wanted to buy fish, but no ponds were ready to be harvested. Customers were reluctant to order STRAL's products if they could not be assured of getting more when they needed them. The partners needed to be able to harvest and sell fish on a year-round basis, says True. "You couldn't go into a grocery store or restaurant and sell them on the idea to put catfish in. They would ask you, 'If I put them on [the shelf] today, can I get them tomorrow?' We didn't have a consistent supply. It took a long time to overcome that—many years."

Farmers were stocking once a year, expecting to grow the fish for a fall harvest. The processing business suffered as a result. "Back then, farmers built ponds as cheap as they could," according to Joe Glover Jr. They "just dammed up a hollow and let the pond fill up from watershed [drainage]. Doing that made the harvesting seasonal, dependent on fall and winter rains to fill these ponds up. It was awfully difficult to establish markets and hold onto those markets, because processors only had fish three months out of the year. They would lose the account because of not having a supply."

The first few years tested their investors' confidence. One investor withdrew from the venture. At times, it appeared as if the processors were not going to make it. Their overhead costs continued, but they often had no operating capital, and running the plant took a tremendous amount of capital. Glover had put everything he had into the business, and the difficulties meant hardship for him and his family. "When my father got out of the grocery store business, he didn't have an income at all," Joe Glover Jr. remembers. "He worked for STRAL for almost a year without any pay, and my mother sewed and made clothes for people. My father, other than being involved in promoting the industry and trying to get it going, used to distribute the *Birmingham News* in Greensboro, Newbern, and Akron." His father had a vision of great possibilities, however, and he stuck with his conviction that the whole enterprise could succeed.

When STRAL had first started processing, catfish were not very widely accepted as food fish, and the general public did not rush to buy them. "We didn't make any money those first few years in the catfish business," says True. "Catfish were scavengers. They were not considered edible fish." The partners knew they needed to make the public aware not only that pond-raised fish were available but that they were good to eat. Glover was certain that, with some effort, a national and even an international market could be found or built. "I knew that there would have to be a tremendous marketing effort put forth, but I had confidence. I knew we had a quality product, one that was liked by people who ate it."

Customers came gradually, with hard work. While Glover and True ran the actual operation, Stephens traveled on the road, often for a week at a time. He introduced himself to managers and owners of restaurants, grocery store chains, and institutional cafeterias. He would return to Greensboro for a short time and then head out to talk to them again or to make new contacts. He communicated constantly the positive aspects of their farm-raised, processed product. After about a year, the processors began receiving substantial orders.

When STRAL did get its first large orders, they came through brokerage houses. The company could not afford to operate trucks to deliver its product to customers directly, so Stephens obtained accounts with distributors. The plant was packaging the fish frozen in the individual plastic ("poly") sleeves and in ten-pound boxes, and brokerages selling frozen foods began to notice the new product. Brokers like Sylvester Chicken in Montgomery purchased fish to sell wholesale, primarily to restaurants. Most of the plant's production went to customers in Birmingham and Nashville like Birmingham Cold Storage, Williams Seafood, and Robert Orr. Those were exciting days, when they began to "net" orders from big distributors such as these.

The STRAL building did not have a very large storage area, so the partners rented space in cold storage facilities. Glover succeeded in making rental arrangements with owners of quick-freeze plants like Robert Gish in Athens and Jake Ledbetter in Cordele, Georgia. They would take fifty thousand pounds of frozen catfish to these warehouses, and the various distributors could go there to check the fish out. The distributors ordered from the cold storage people, who sent records of the orders to STRAL. The buyers then paid STRAL directly. In effect, these distributors were selling the fish for the partners.

Some of STRAL's fish ended up in restaurants, and some ended up in grocery stores like Piggly Wiggly. Getting the fish onto shelves at Piggly Wiggly and other stores, however, often meant traveling and selling them to every store in the chain. Stephens would go to Piggly Wiggly's central warehouse, talk to the buyer about catfish, and offer to collect orders from each individual store himself. After obtaining agreement from the buyer, he, and sometimes Glover, would drive to every town where there was a store. They would go in and show the manager a flyer about their fish, give them a sample of the fish, and get an order. Then they would take all the orders they had collected and give them to the buyer at the central warehouse. The buyer ordered 100 or 150 cases of fish and distributed them to the stores. So basically, the partners had to sell their product from scratch at each individual store. When the individual stores got ready to reorder, however, STRAL had a foot in the door. The stores' managers reordered directly from STRAL.

The processors were more fortunate with Winn-Dixie. Next to land that Stephens owned was a cattle ranch belonging to Fred Davis and his son. Davis had worked with Winn-Dixie as part of his job, and when Stephens approached him with news of catfish for sale, Davis talked with business acquaintances at the store. That neighborly connection proved very helpful to Stephens in obtaining an account with the supermarket chain.

One cool, fall afternoon in 1968, shortly after they had moved into the new building, Glover and Stephens headed over to Selma with trucks and a crew to harvest a pond they had stocked earlier that year. The two men, the pond owner, and the crew put on their waders and went out into the water to seine the fish. A musty, sort of moldy smell hung over the water.

As they were pulling the fish in, Stephens asked Glover, "Say, Joe, what do you suppose that smell is?"

"I don't know," responded Glover, equally puzzled.

No one knew what the smell was, but they continued the harvest as planned. The pond yielded several thousand pounds of nice, healthy-looking fish that were taken to the plant, processed, and packaged. As usual, the boxes were marked with a code to identify the batch and placed in the freezer.

It so happened that around that time, the Quaker Oats Company had been trying to enter the catfish processing business. A representative from the company's mill in Chattanooga had visited the STRAL plant to discuss the possibility of buying their business. He had been there on the same day that the fish from the odd-smelling pond were being processed and had asked if he could take some back with him to sample. Later, after all the fish had been packaged, he came by the plant again to pick up some boxes before driving back to Tennessee.

Around nine o'clock that evening, Stephens's phone rang at home.

"What in the world is wrong with those fish you sent?" demanded the man on the other end of the line. "They're the smelliest fish in the world! I just cooked some up for supper, and my whole house smells like it's been fumigated! We can't eat them! They're terrible!"

Stephens did not understand what the man was talking about. "What do you mean? What's the matter with them?"

"I haven't any idea," answered the man. "We've cooked two or three batches. We opened several boxes. They're terrible—the worst-smelling things I've ever had! Something's *wrong* with them, and that's the truth, by George. *You* go try 'em!"

Stephens obtained from the Quaker Oats man the numbers that had been stamped on the box. This was too important to wait until morning. He left the house and drove to the plant in the dark of night. He went into the freezer, found a box of fish from the same batch, and brought them home. His wife, Minnie Steele, cooked them, and sure enough, they were awful. It was the worst taste he could remember ever experiencing.

The processors had no idea what was causing the terrible taste. The rest of the boxes in that batch of fish just sat in the plant freezer. The partners could not even give them away. Stephens felt obligated to pay the farmer, since they had taken his fish, and he did just that. The ten thousand dollars they had agreed to pay for the fish took a large bite out of STRAL's cash flow. The harvest was a dead loss for them.

This was the partners' first experience with off-flavor. After that day late in 1968, they checked the fish in every pond before they harvested them.

Two or three months later, a farmer friend of Stephens asked STRAL to come get his fish. "They're ready for you to take," he told the processors.

Someone headed from the plant over to his pond to sample the fish. The testers came back and reported that the fish were off-flavor. Not without difficulty, Stephens told the farmer that STRAL would not be able to take the fish.

"What do you mean? There's nothing wrong with these fish. They're good!" the pond owner protested. "Why, we've eaten them ourselves!"

Stephens did not waiver. STRAL had already lost too much in its first encounter with off-flavor. "I'm sorry, but we can't give our customers those fish," he said.

The farmer became angry. He needed to sell his fish to recoup the money he had put into them. Things came close to turning ugly that day. The farmer calmed down eventually, but he did not give up.

A taste test one week later gave the same result. L. C. Alsobrook, a county extension agent, tasted the fish and knew they were not right. With his help, the processors contacted the fisheries program at AU, where researchers had just started to examine the problem of off-flavor. They were told someone from Auburn would come to Greensboro the next day to help. In the morning, Oneal Smitherman arrived at the plant. The men drove to the pond. Smitherman cooked a fish while they were still on the pond bank. The fish smelled terrible. He shook his head and declared them off-flavor.

The farmer was enraged, but no amount of objecting made the STRAL partners go back on the decision. Somehow, Smitherman and the others present helped calm the farmer down. He agreed to another taste test to be carried out after some time had passed.

A few weeks later, the farmer and his wife came to Stephens' house for dinner, bringing with them a package of their fish. They cooked them right there in Stephens's kitchen to show him that the fish were perfectly all right. They sat down to eat. The smell was not good.

The farmer and his wife ate and declared with satisfaction, "Now, that's good fish! Nothing wrong with those fish!"

"They stink," came Stephen's response.

"Y'all just don't want to buy them!" accused his farmer friend. The feeling in the room was extremely awkward and unpleasant. The evening finally came to an end, but nothing had changed.

Over the next several weeks, two or three heavy rains soaked West Alabama. STRAL operators kept checking the farmer's fish. One day, the musty flavor was gone. After all the trouble everyone had gone through, waiting those weeks had cleared up the taste, and STRAL could finally buy the fish.

7
Fried Chicken of the River

> In 1968–69 there were a number of ill-fated attempts to establish fast food restaurants specializing in catfish. The majority ended in failure.
>
> —Gary C. Cox

The STRAL Company appeared to be operating successfully during 1969, even though it was still barely managing financially. The hatchery pretty much controlled the catfish fingerling market for miles around. The plant was the only one processing mature fish grown in the area. The partners had been able to keep a fair-size inventory of fish on hand to sell. What STRAL needed most was more sales. The business was heading toward a vertical integration typical of later, much larger companies.[1] "What I envisioned," says Check Stephens, "was a little bit ahead of our time, and the timing in these things makes a big difference. The ideal operation . . . is to hatch the fish, grow the fish a certain percent of the time by yourself, have the plant to process them, and then sell the fish. You sell the fish through your own operations, like restaurants, for the maximum amount of money. That's the way I was trying to build this operation we had."

Richard True and Stephens had been stocking fingerlings for some time in Charles Dixon's ponds in Andalusia, in south Alabama. On one of their delivery trips in 1968, the conversation with Dixon turned to the idea of establishing a catfish restaurant together. Minnie Pearl, the country singer, had set up Minnie Pearl's Fried Chicken, and Colonel Sanders' Kentucky Fried Chicken had become popular. C. A. Ezell's catfish camp was well liked. Why not a fast food catfish restaurant? Dixon and John Vick, his son-in-law, could build the restaurants, and STRAL could be a partner in the business and provide the fish.

Early one day in 1969, while True and Stephens were busy at the hatchery in Newbern, a car rolled up the driveway. A man got out and walked over to them. He stretched out his hand in greeting.

"Howdy. I'm J. B. Smith," he introduced himself, smiling.

The men shook hands.

"J. B. 'Initials Only' Smith," he went on. "Middle initial 'B' is just that—initials only. When I was in the military, they used to call my name out: 'J. B. Initials Only Smith.' Guess it stuck," he explained with a grin. "I understand y'all are in the catfish business here. Well now, I've got a Shoney's restaurant in Tuscaloosa. I'm putting something together, and we're going to need a heap of fish. I'm thinking I'd like to get them from you fellas." Smith had come down to talk about supplying catfish for a group of restaurants that he and some partners in Nashville, Tennessee, were planning.

After the exploratory discussions Stephens and True had had with Charles Dixon, this looked like it might be a good opportunity. The STRAL partners talked about the possibility with Dixon and his son-in-law, who agreed to give it a try. They reported back to Smith that they were interested.

"Well now, that's just fine," Smith said.

Numerous conversations with Smith followed. The prospect looked better and better, and they started getting things ready for a formal agreement. It was then that unease began to creep up on the STRAL partners. They still had not met their partners-to-be and began to hesitate, thinking they should perhaps not get involved in an agreement with people they did not know. When Dixon talked about backing out, they let the whole matter drop.

Smith, however, did not. He came back to Greensboro with friends from Nashville—Dennis Wright, a lawyer, and Frank Davis, who owned a Buick dealership and a franchise on a number of McDonald's hamburger restaurants in the Nashville area. Davis was also the chairman of the board and president of the corporation that had been started for this enterprise. They wanted to establish a franchise catfish restaurant chain under the name "Catfish Shanty." Wright had been in touch with some people in Nashville, Grand Ol' Opry people, who were interested in buying stock in it. This time, the three men had come to ask them to supply the fish. STRAL was the only company processing catfish on a significant scale at that particular time, so the setup began looking good again to the STRAL partners.

J. B. "Initials Only" Smith was intent on making the deal work. He stayed in touch with the processors and informed them that others were part of the venture also, namely Minnie Pearl; Dizzy Dean, the baseball star; and Pat Dean, Dizzy's wife. Hence the name of the corporation: Miz Diz Systems. Things certainly looked good. If everything worked, True and Stephens would have a big market for catfish, just as they had wished. They agreed to buy stock in the venture.

Not long after, Stephens received a call from Dennis Wright. Wright wanted Stephens to come to his office in Nashville to meet with him. Stephens went. All day long, Smith and Wright tried in various ways to convince Stephens to sign papers giving them exclusive rights to the processing company's total production. "We'd like fifty thousand pounds," they said, "but we'll take whatever you've got." All day long, Stephens refused. He returned to Greensboro, shaking his head. He talked first with True and then with Charles Dixon, who had originally agreed to finance the restaurant end of an enterprise with STRAL.

"They really put me in a hot box up there," Stephens told Dixon. "I just don't like the way they want to set things up. It doesn't sit right with me. They know we're the only people around who are processing catfish, and they were wanting my signature so they could have a lock on our business."

Dixon agreed. "Well, let's just not do it, then," he declared.

Stephens nodded in agreement. Case closed. Or so they thought.

At about the same time, STRAL had a large amount of frozen fish in storage that needed to be sold. Approximately a hundred thousand pounds were sitting in a Birmingham warehouse, and close to fifty thousand pounds were filling up the plant's own coolers. The business needed money badly. The partners were not making enough to continue paying their employees and the rent on the warehouse, much less any of STRAL's stockholders. Neither Glover, True, nor Stephens had any money themselves to put into the company. They had to do something.

Frank Davis seemed to be an approachable person. Stephens got in touch with him to see whether, if Miz Diz was still interested, Davis might be able to help STRAL obtain a reasonable contract. While waiting to hear back from Davis, Stephens and True left to deliver two truckloads of fingerlings in Monmouth, Illinois. On the way, they stopped to call home for messages.

"Frank Davis is trying to get hold of you," Stephens's wife told them. "He wants to talk to you."

Stephens called Davis back right away.

"Let's meet," Davis said. "I want to talk with you right away. Can you meet us down at Dizzy's house?"

Dizzy Dean's house was just outside of McComb, Mississippi. Stephens and True agreed to go there on the trip back from Illinois. Stephens called Edgar Russell, their lawyer, and asked him to join them for the meeting. Frank was going to be down there with his lawyer. STRAL ought to have its attorney there too.

Stephens and True unloaded the fish in Monmouth and then turned the empty trucks around and drove southward all night long to Laurel, Mississippi, where they met up with Edgar Russell. The two men had had no sleep, which was par for the course. Leaving the trucks at the hotel, Stephens and True rode to Dean's house in Russell's car, sleeping almost the entire way.

Frank Davis and his attorney had already arrived at the Deans' large plantation home. "Miz Diz" and Dizzy Dean were there. They shook hands all around. Everyone was invited into the dining room. Dizzy Dean had to stoop to walk through the doorway, he was so tall.

A delicious meal had been prepared for them. The conversation kept revolving around STRAL's production. Davis and Dean tried to get STRAL to commit one hundred thousand pounds of its production exclusively for Miz Diz Systems. They offered the men a substantial amount of stock in the corporation, but the STRAL partners and their attorney hesitated. Oddly, during the entire meeting, no one had mentioned anything about signing a contract. Edgar Russell decided to make something happen.

"Well, we've enjoyed the meeting," he said, pushing his chair back from the table. "I don't guess we have anything to talk about." He rose to leave. Stephens and True took the cue and moved to get up also.

Immediately, Davis and his attorney called them back. "Hey, now, just a minute! Wait a minute now!"

A contract was produced and signed. That evening, when Russell, True, and Stephens got back to Laurel, they slept the night in a hotel, satisfied that they had held their own and come out of the meeting with a contract that was very good for STRAL Processing Company. STRAL had agreed to send catfish to Miz Diz Systems to be served in the Catfish Shanty restaurants.

~

"Good morning, Richard. Have you read this?" Check Stephens greeted Richard True on the morning of July 14, 1969, and handed him a folded newspaper—the business section of the *Selma Times-Journal*. True unfolded it and noticed the headline on the first page. It read: "Catfish May Be Ready to Give Chicken Run." After glancing at the photograph of Check and himself standing next to one of their hatchery troughs with a local county extension agent, his eyes dropped down to the lead paragraph of the article: "*Greensboro*—'Miz Diz Catfish Shanty' may soon become as popular as other chain restaurants throughout the South. But this one is different from most. It specializes in channel catfish and hush puppies."

True read on: "Raising catfish is going commercial throughout the South and may soon give chicken a good run for the consumer's money. Total poundage produced this year in four states—Alabama, Arkansas, Mississippi and Louisiana—is expected to be about 70 million pounds. Some people have already tagged this new product as 'the fried chicken of the river.' Behind the drive are such producers as Check Stephens and Richard True of Hale County. [The latest] development is a chain of franchise catfish restaurants—Miz Diz Catfish Shanty, named after Dizzy Dean's wife. Sixty franchises have already been sold throughout the South."

The arrangement worked out well for STRAL. Miz Diz bought fifty thousand pounds of fish, the largest single sale STRAL had ever had. The selling price was terrific, and, though it took a while, they eventually received all of what was due them for the fish. They were able to pay off all their debts, with enough left over to leave them in good financial shape.

Fish moved steadily through the plant. They were shipped north to Nashville, where they sold like hotcakes in the Catfish Shanties. A restaurant was established in Birmingham and another in Montgomery, and plans were afloat for others.

The restaurants had been operating for a year or so when A. H. "Pete" Crow was hired to run the restaurant operation. Crow was executive vice president of McDonald's restaurants at the time. One day not long after he was hired, Crow had a talk with Stephens. That talk sealed the fate of STRAL's involvement in Miz Diz Systems.

"Check, I'm in the merchandising business," Crow said. "I know what works and what doesn't work in this business, and one thing we've got to do is start having strict portion control," he said. "What we need to do is have the fish as whole fish."

"What do you mean, whole fish?" asked Stephens becoming concerned.

"I want them to be 5- to 7-ounce, 7- to 9-ounce, and 9- to 11-ounce fish."

Stephens balked. STRAL's fish were harvested annually, after growing to around one pound in size. He shook his head. "Pete, there's no way we can supply that."

"Well, you've *got* to supply it. That's what we want to do," Crow insisted.

"Pete, we can't do it. The only way we can supply you is with the big fish. It's going to take an awful lot more fish to provide what you'd need, and we just don't have the capacity to grow enough of those little ones. We just can't supply you," Stephens insisted in return. Crow was determined, however, and Stephens's arguments did not prevail.

STRAL tried but was unable to supply the Catfish Shanties with enough fish in the desired sizes. Crow knew the restaurant business, but no one could supply him with the fish he wanted. They were just not available at the time, because no one was producing and processing them in those sizes and quantities. The corporation had already begun offering stock to raise money to pay for the growing chain, but the economic situation took a downturn. A number of short-order restaurants that had recently been established started to fail, including Minnie Pearl's Fried Chicken. Investors were short on capital and did not want to risk any on Miz Diz Systems. The Catfish Shanties folded.

The Catfish Shanty was STRAL's first venture into the restaurant business. The processors were fortunate to get all of their money out of Miz Diz Systems before it went completely under. "Timing was the big thing," comments Richard True on the restaurant affair, "because several have come on line since then and have done very well. We were just premature."

Notes

1. Companies that are vertically integrated engage in and control more than one phase or step of the processes on which their business relies.

8
From STRAL to Farm Fresh

> We were used to having small banks, small grocery stores. Then they put a supermarket in, and a little store would go out. All the people interested in the fish business—the major people with the unlimited finances—it became quite apparent that, as time went on, we'd have a hard time competing with them. You'd almost have to rob Peter to pay Paul, wondering where you were going to get the money to meet the payroll.
>
> —Check Stephens

When STRAL Processing Company was operating in the back of Glover's Grocery, most catfish being raised in the area were sold live at pondside. After the new building was built, the partners agreed to harvest and buy from more farmers. In the fall, fish came in faster than they could be sold, and the inventory built up. The men knew that supplies would wane in spring and summer, so they froze the fish for later sale. "We were buying them, trying to keep the farmers happy and continuing to raise fish," Richard True recalls. "We bought a lot in October, November, December, and January. We'd freeze and stockpile them and then spread them out over the year." Stockpiling thousands of frozen fish meant that the large amounts of money spent on purchasing, harvesting, processing, packaging, and storing them was not being recouped.

By 1969 production and processing enterprises were popping up around the South, especially in Mississippi. STRAL Processing did not have the support or financing that Mississippi fish farmers did. The partners were not able to maintain their market and make their business grow. "We were smothered with frozen fish," says True. "We just couldn't handle it."

In spite of the ups and downs the industry was experiencing in the late 1960s, however, optimism about its long-term potential generally prevailed. The small, pioneer STRAL venture looked very attractive to larger, established companies involved in other agricultural processing activities. Within less than a year after being established, STRAL Processing was receiving offers from businesses with significantly greater resources than its own. Companies involved in feed or poultry production or processing were now eyeing catfish. "Big Henry" Williams came representing Goldkist, which was already involved in poultry processing, cotton, and other products. He and the STRAL processors did not come to an agreement. In 1969–70 Goldkist decided instead to build a small catfish processing plant at Quitman, Georgia, to process catfish raised in East Alabama and West Georgia. Not enough fish were available in that area, however, and the plant closed down after operating only a short while.

Other companies made similar overtures to STRAL. Purina officials flew down to Greensboro in their "checkerboard" airplane and spent a day with the STRAL partners. They looked the operation over and talked about becoming partners but decided instead to start their own plant in Mississippi. The Alabama Farmers' Cooperative considered purchasing STRAL but decided to locate elsewhere also. Representatives from the Quaker Oats Company came down and offered to buy the entire business. The STRAL processors discussed a possible partnership with Ralph Wells Company, but that company would agree only to a total purchase.

Nebraska Consolidated Mills made an offer to merge with the smaller Greensboro business. Stephens did most of the negotiating, and haggling about various points went on for weeks. Finally, the two parties came to a mutually acceptable arrangement. In 1969 the original STRAL catfish processing plant and hatchery became part of Nebraska Consolidated Mills, later to become ConAgra. Stephens, True, and Joe Glover had hoped to avoid selling the entire operation and have some kind of joint ownership arrangement. They had not originally intended to sell out to ConAgra, but, according to Stephens, "that's what we *had* to do. It was the right decision. There's no question in my mind we would have wound up down the tube, the way things were going. It was going to take a pot full of money to really do anything with this fish business." In one way it was a relief to True. At least, they figured, they would now be able to afford to improve the business. "We now had the resources to do something. I don't say we had a lot of money, but we had the knowledge that the capital was there if we wanted it, if we had something that needed to be changed and looked like it was a reasonable idea."

As in many small, rural towns, there was a certain amount of resistance in the community to the sale of STRAL to an outside company. "Lots of people were diametrically opposed to us merging with ConAgra," Stephens recalls, "because they didn't like to see outsiders. They had hoped it could have been kept local and all, but you needed to have way more than what we could put in it."

The three men agreed to stay and act as a management team under Sam Hinote, who came to oversee operations. True had a seven-year contract to manage the hatchery, and Stephens, working under a similar contract as general manager, coordinated business at the plant and the hatchery. The company expanded the plant, and Glover managed and supervised everyday operations. He did not have a contract, but he continued on in the hope that the new financial strength from Nebraska Consolidated Mills would allow the business to make improvements he had been unable to make at STRAL for lack of money. Nebraska Consolidated Mills kept the STRAL name and continued to operate with one processing line. After the purchase, the number of employees increased significantly. The hatchery employed several people; the processing line employed about thirty; and the harvest crew, about five. By this time, STRAL had large delivery trucks and four or five full-time drivers who delivered either fingerlings to farmers or fresh or frozen fish to customers.

Check Stephens's career with Nebraska Consolidated Mills was short-lived. After about a year, he decided to leave to work for himself. He and the company worked out an agreement. He purchased some of the STRAL broodstock he had helped develop and eventually started his own hatchery and fingerling business in Selma.

Joe Glover stayed on with Nebraska Consolidated Mills for only about a year, too. He had hoped he would be able to upgrade and mechanize various aspects of the processing, but getting support for his ideas from upper management was a struggle. The plant did not produce as well as he felt it could. Although he was convinced that the continued success and growth of the processing industry depended on obtaining a year-round supply of fish, the company kept the plant going on a seasonal supply base. In spite of the company's financial strength, the business did not grow.

Glover began to think increasingly about operating his own company. He was convinced he would be able to move fish faster and make a larger profit. He opted to leave in 1970 and built his own plant that year right across the street from STRAL. He named his operation Farm Fresh Catfish Company. A few years later, ConAgra would sell the STRAL plant to a group of local farmers and move to a processing operation it had established in Jackson, Mississippi. True, the only one of the three original STRAL partners to stay with the company, would move to Mississippi to co-manage with Sam Hinote the company's three plants there.

When the twenty-or-so Farm Fresh plant employees showed up and started to process fish in November 1970, Nebraska Consolidated Mills suddenly had competition. At that time of year, catfish farmers everywhere were ready to sell their grown fish. Greensboro-area growers were happy to have two processing plants nearby. They would now be able to sell their fish to the highest bidder rather than have to settle for whatever price they were offered. The two plants and the two former partners, True and Glover, found themselves competing for fish in the same market.

That competition resulted in problems for both the farmers and the processors, especially in the fall, when all the farmers' fish were ready for harvest. So many fish were available that the processors could afford to offer lower prices for them. The farmers, however, were naturally looking for the best offer, so they would play one processing house against the other as prices changed. As a result, True would often find himself wondering where he was going to get the fish the plant needed. "By November, everybody had fish for sale. The plants didn't need them," he recalls, yet "it was a balancing act trying to keep the farmers happy so you could have fish later on. What usually happened was, the farmers selling me fish became disenchanted with me for about two years. And every two years they would be at the other plant, then head for my plant. That's how it went, a constant swap."

A surplus of catfish may have existed in the fall and early winter, but spring and summer brought a short supply. Farmers generally restocked their ponds in the spring, because those that had been drained for the previous harvest took time to fill up. The fingerlings needed the warmer months that followed to grow out to a size processors could use. Because Farm Fresh could not find enough fish during that part of the year, Glover had to look elsewhere. According to his son, Joe Jr., not a year went by when his father did not have to haul some fish from Mississippi to supplement his inventory.

The Farm Fresh harvest crew did all the harvesting, traveling sometimes hundreds of miles and toiling long, exhausting hours to "scratch out" what fish they could so the plant could continue operating. The bulk of the work was sheer physical labor, taking fish from ponds that were poorly designed for the use of seines. Glover and three or four other men would take the harvest truck, nets, generators, and other equipment. Often, the farmer himself would help (see fig. 9). "We went as far as Bay Minette south and to

Fig. 9. Farm Fresh harvest at Earnest Broughton's pond in Greensboro, early 1970s. On the truck, Joe Glover (white shirt) and pond owner Earnest Broughton are weighing basketloads of fish brought up from the water. In back of the truck, Broughton's wife, Alieze (white hair), is recording the poundages. In front of her stands Gaither Thomas (overalls and cap), a neighbor and kinsman of the Broughtons. Other individuals pictured are unknown. (Photo courtesy of USDA, Soil Conservation Service, now Natural Resource Conservation Service.)

Sand Mountain north to harvest, or attempt to harvest. A lot of little ponds were harvestable, some you just could not seine. It was a nightmare. Some farmers had built their ponds with a small drain in them, maybe not even a drain at all, and you couldn't get the water out. We would go one day and have to wait for the water to drain through a four-inch pipe all day and all night and maybe catch a few fish the next day."

~

Late one morning on a warm September day in 1971, the Farm Fresh crew finished stowing equipment after a harvest. Joe Glover and the farmer finished talking and shook hands. The harvesters gave their faces and necks a last long wipe in the hot sun and climbed into the cab. Glover turned the refrigerated truck away from the ponds, and it lumbered slowly down the dirt road toward the highway. Greensboro and Farm Fresh were miles away.

It felt good to be finished with the sticky mud and the heavy dragging and lifting. The men stretched out their legs as best they could and savored the warm, humid wind that blew over them. Dirty and sweaty, tired and thirsty, but temporarily relieved of their hard work, they rode for a while in silence. Later, they recounted some of the more difficult moments of the job and teased each other about more amusing ones.

A couple of hours later, the truck rolled into the Farm Fresh lot and backed up to a rectangular holding tank at the rear of the plant. Mustering up the energy to start working again in the heat, the men slid down from the cab. Among them was Randy McCrory, a young fellow who did a variety of jobs at the plant and who had worked on the harvest crew that day. He walked around to the rear of the truck and unlatched the big doors. As he opened them wide and stepped aside, a torrent of fish flowed out and splashed into the water waiting below. After a few moments, the torrent slowed to a trickle and then stopped altogether. There was no way to hold water in the truck to haul the fish live, and most had suffocated on the trip. If the crew had taken Joe Glover's flatbed with the tank, it would have been much easier to get the fish out, but that truck could hold harvests of at most only about two thousand pounds.

Taking a large, old push broom, Randy clambered up into the trailer and set about sweeping out the fish that remained. He could not avoid walking on the layer of fish still covering the floor. Although their dorsal spines now lay flat against their backs, he still had to be careful not to slip and fall and slide into them. He pushed the fish with his broom toward the rear opening till the only fish that remained were the ones stuck to the ribbed metal floor. The bream always seemed to stick. When they needed fish for the plant as much as they had that summer, they often had to agree to harvest ponds that contained "trash" fish—species other than catfish, like bream, which were, for one reason or another, not marketable.

On this particular trip, the men had not had time to place cardboard on the trailer floor first. Without the cardboard, cleaning the floor was much more difficult, but a large order of processed fish was scheduled for delivery to Birmingham that day, and Farm Fresh had no other trucks for deliveries that size. The truck needed to be cleared and cleaned as soon as possible. So Randy pushed, pried, and scraped, until he had swept all the fish off the trailer floor and into the tank outside.

In the meantime, other workers had been busy spreading the pile of fish out in the holding tank. On either side of the tank, two men were leaning against the crowder bars to push the bars along on the rollers. The bars, which extended across the tank, were designed to force the fish through the water toward the other end. At first, however, the huge pile was a solid, resistant mass. Veins stood out on the men's foreheads and arms as they leaned and pushed. Pushing thousands of pounds of catfish by hand was hard work, but eventually, the mass of fish moved more easily, and the men gathered the fish together at the other end of the tank, closer to the building. They had to get the fish through the processing line before lunch time for the load to be packaged fresh before the plant closed at the end of the day.

The day before, the harvest crew had brought fish from ponds closer to Greensboro, and a number of the fish had survived the ride. Those had had to be stunned before they went into the plant. Fish flopping around in the hands of the band saw operators posed a danger to the men operating the machines. Because stunning the fish involved sending electrical current through the water, shocking them was itself a dangerous job, and in later years, the practice would be stopped. On that particular day, however, Randy had taken the battery cable terminal and clamped it into an outlet on the tank. Because it was summertime and the water was warm, it had worked instantly, and the stunned fish had risen almost immediately to the surface.

Today, however, the fish had not survived the trip, so they did not have to be shocked. The men picked up handbaskets and began dipping fish out of the tank and tossing them into a chute that sloped down through an opening in the outside wall of the plant. The men were able to lift between thirty and fifty pounds with each dip, quite a feat for a job done by hand.

The load that day amounted to about six thousand pounds. The plant could handle ten or twelve thousand pounds a day, so if the plant workers could keep the line moving, they would be able to finish before the day was out.

The fish came down the chute into the building, and the processing began. Everyone soon fell into a steady rhythm of work. The fish slid down onto a long, traylike table inside. Two men inside at the bottom of the chute took them one by one onto the band saw tables and pushed them through the blades, cutting off the heads. Fish heads fell steadily into the fifty-five-gallon drum waiting under the edge of the saw tables. In later years, a conveyor belt would carry the heads away, but for now, when that barrel was full, a man closed it up, replaced it with an empty one, and rolled the sealed barrel out to another room. The men outside continued to scoop fish out of the tank and send them down.

Once the heads were off, the fish were placed onto an adjacent, sloping stainless steel table down which they slid, stopping gradually at a second table. There, two women on either side stood ready with sharp knives. One by one, the ladies drew each fish close and with a quick, clean slice, slit open its underside. From there, the split fish were passed to two other women, who reached into the bellies and pulled out the entrails. With their bloody gloves, the women pushed the entrails through a grate in the table into a metal washtub below. Periodically, a man came over and emptied the contents of the full tubs into fifty-five-gallon barrels. The eviscerated fish were placed on a large tray next to two nearby skinning machines that stood on either side of the processing line.

Joe Glover Jr. was one of the two men stationed at each of the skinning machines that day. Joey, as friends and family called him, had learned about growing catfish since 1965, when he had helped with his father's first pond. He had seen firsthand how STRAL had gotten its start, watching at his father's store in the afternoons after his high school classes let out. After graduating from high school and spending two years in the military, he had started college. Undecided yet as to what to do after college, he worked at the plant during the summer. He wanted to earn some money and see what the processing business was all about. Sometimes he worked on the harvest crew; sometimes he helped on the line. He was already adept at deheading catfish on the band saw. Today, he was honing his skills at skinning. An experienced operator had shown to him how to hold the fish to the machine so the skin would come off cleanly. Joey did not want to hold up the line. His performance on the job showed that he had learned from his father the importance of hard work and of doing his level best.

After the skins were removed, fish were sent sliding down to another steel table where two women were washing all the fish that came their way. When Farm Fresh had first started, women had washed each one by hand under a stream of water pouring into a large vat, just as they had at the STRAL plant. Joe Glover had had the idea that electrically powered brushes and running water would clean the fish out much more quickly and efficiently than could be done by hand. Recently, he had devised and installed such a special cleaning machine. Now, the women held each fish over the sink and, opening up the belly, held it up to the soft brushes that turned on the roller. The water streaming out helped clean the fish. At that time, before the days when latex gloves were used, processing line workers wore cotton gloves, and the cleaners' gloves were soaked with water and blood. The wetness felt refreshing in the hot afternoon. Once both the inside and outside were rinsed clean, the women laid the fish on another small table to the side. Others determined the size of each one by placing it on a scale or estimating by sight and then packed it into a long, narrow plastic bag that they laid in a large plastic pan. A man brought ice with which to cover the fish and keep them cold as they were put into the cooler. Within a few years, Glover would have an icehouse installed at the end of the table, one that would make ice continuously for the expanding volume of fish going through the line.

Sometimes, especially on hot summer afternoons, a distant, occasional rumble of thunder would threaten. Summer afternoons were a far cry from what the plant was like in the winter, when everyone had to wear extra layers of clothing and always wanted to have the big heaters turned on. Suddenly, the machines stopped, and the lights went out in the big room. The power had gone out—a fairly common problem in those days, particularly during thunderstorms.

Equipment sat idle, partially processed fish lay around, and not much could be done about it. Someone brought buckets of ice and poured it over the fish lying on the tables to keep them chilled down. Most of the time was spent just sitting, talking, and waiting. The break was welcome to the workers, whose jobs kept them standing for hours on the concrete floor. Finally, after some minutes—sometimes as much as half an hour—the lights would come back on and the machines begin humming again. The fish were processed and boxes were loaded into the freshly cleaned, refrigerated truck before nightfall. Today, unlike some other occasions, they did not have to work late into the night to get the order out in time for the scheduled delivery.

~

Joe Glover Sr. was constantly trying to thinking of ways to make the Farm Fresh operation work more smoothly, safely, and efficiently. He read tirelessly through trade journals from other food processing industries. "We always got ideas from some other industry," according to Glover. "Other machines had already been put in by the meat packing industry, not from within the fish industry. We had to go outside. When you're in a new business like that, you read everything you can to find out how other people handle their products." Little by little, he incorporated new ideas to "give a smoother flow through the plant, less handling and manual work, better grading, better freezing, better storage." The entire system improved, one step at a time. Glover explains, "At times we spent money that was the wrong move, but you had to take chances. It worked, but maybe not as good as we'd like, so we redesigned it. It was a gradual process. You'd get a little improvement this year, a little bit more next year, a little more the next. We just kept working at it."

In 1971 Glover installed a grading machine on the line. Hanging metal bowls swung about a foot apart along an oval track above the weighing table. Each bowl hung from its own scale. Workers lifted a fish into each bowl as it swung by. Grading by weight, the bowls were calibrated to turn over and empty their load into various tubs. Six-ounce fish fell into one tub; seven- to eight-ounce fish into another, and so on. Workers slid each graded fish into a plastic bag and laid the bags in ice-pack boxes. Once full, the labeled boxes were closed and placed on basket trays like the ones on which bread is still transported today. Stacked onto carts, the trays were rolled into the freezer.

Years later, when Farm Fresh was well established, Glover had a liquid nitrogen freeze tunnel installed, with conveyor belts on which workers would lay the raw product. The belts would carry the fish or filets through the quick-freezing unit and deliver them frozen at the other end. There, people would grade them and put them into boxes. The quick-freezing liquid nitrogen process was expensive to use, however, so Glover later installed a spiral freezer. The mechanical spiral freezer had much greater capacity and was much less expensive to operate when large numbers of fish were being processed.

Because of the glut of fish in the fall, catfish was not considered a viable retail item at that time of year. At Thanksgiving and around the Christmas and New Year's holidays, grocery stores wanted to save their valuable cooler and freezer space for hams and turkeys and other traditional foods that sold well during the holiday season. Farm Fresh could not sell all the fish it had processed. But catfish were ready to harvest, and the farmers wanted processors to take them.

Glover had already realized that having a seasonal supply was not going to work in the long run. Restaurants and retailers wanted a product that was available all the time and not just a few months out of the year. In order to sell enough catfish to succeed as a business, Farm Fresh had to be able to offer its products to its customers on a regular basis. Seasonal harvests would have to be eliminated. But to farmers, catfish were as much a seasonal crop as wheat or soybeans. Glover knew that fundamental changes would have to be made. He knew he would have to get farmers to stagger their stocking so there would be fish maturing at various times throughout the year.

Convincing the farmers was not going to be easy. Glover took time to talk with them about it. He invited them to come sit in his office and discussed the problem with them. He explained that without a year-round supply, the industry could not succeed. He explained how their late fall supply did not fit well with the wishes of consumers, who wanted the fish all year long and particularly in the spring and summer months.

Glover was asking farmers to think about catfish differently from their other farm crops. Until now, farmers had always felt that they had to sell their crops off as soon as they were ready for harvest. Processing catfish was not the same as ginning cotton, however. Catfish could not be processed in the fall and winter and then kept in freezers indefinitely, the way cotton could be kept in warehouses. The fish became rancid. They got freezer burn. When that happened, the crop was a total loss.

Glover recognized that farmers needed an incentive to change. As part of his campaign to make the industry viable, he began to pay a slightly higher price for fish in spring and summer, when the supply was lower. He hoped that farmers would respond by staggering their stocking. However, the idea was still too new to them. Nearly all resisted the change. "They didn't like the idea," Glover explains. "They just had to get used to it and learn that it could be done. It takes eight to ten months of warm-weather feeding to grow catfish to a marketable size. In order to have fish year-round, they had to be stocked every month of the year. It takes longer if you stagger, because they do not grow much in the winter months. So you need to stock them in July of this year for them to be ready for market in July of next year."

At first, only a few of the farmers understood Farm Fresh's marketing needs and agreed to grow catfish throughout the year. Glover persisted in talking and working patiently with those who did not. Eventually, they too saw the connection between the trouble they had selling their fish and the trouble processors had with sales in the fall. In contrast, during spring and summer months when demand was stronger, the farmers knew their fish sold well. Through the years, Glover retained his quiet conviction, and although it seemed sometimes as if the plant and the business might fail, he did not give up.

The activity at the plant belied the difficulties of keeping Farm Fresh afloat the first year. Sales just did not keep up with production. Before long, Joe Glover was unable to meet his payroll obligations to his employees. He had to do something very soon.

The best chance he had at the time was to turn to the producers for help.

Clearly, catfish farmers had a stake in the future of a local processing enterprise. Surely they would be willing to invest in Farm Fresh. Glover was convinced his hunch about the future of the industry was correct. Much as he wished he did not have to do it, he paid visits to the wealthier farmers he knew and asked them to buy stock in his company. He promised to buy the stock back as soon as the company could afford to do so. Because the farmers knew him and knew that his word was good, he was able to find ten who agreed to purchase five thousand dollars worth of stock each. Glover now had a total of fifty thousand dollars to meet the plant's needs until sales picked up.

The borrowed money ran out within a year. Well into 1971 Farm Fresh still did not have an adequate cash flow to pay for operating expenses, much less make a profit. Glover turned to a local banker, Lewis Lawson, for help in getting a loan from the Small Business Administration. In order to get the loan, Lawson needed collateral. What Glover was able to offer was inadequate to cover the large amounts needed to keep the plant operating. But Lawson knew Joe Glover to be a hard worker and a man of integrity, and he decided to take a risk for him. Lawson guaranteed the loan himself. That was the turning point for Farm Fresh and for Glover's dream. After he got the loan, he was able to begin eking out a profit and pull the struggling business through.

One morning in 1974 or 1975 Randy McCrory knew that something was afoot when Mr. Glover asked him to get several tables. They were standing in the doorway of a small room Glover had added at the back of the plant, just off the main processing area.

"What are we gonna do, Mr. Glover?" Randy asked.

"We're going to filet fish, Randy," answered Glover.

"Yes, sir," responded Randy as he turned to round up some tables.

McCrory was full of curiosity. He had no idea how to filet a fish. Men set up a band saw in the little room. Glover gathered Randy, some of the women from the line, and two or three more people he had recently hired into the room.

"Randy, would you bring us a bucket of fish?" he asked. Randy obliged.

Glover turned the band saw on and picked up one of the fish. Holding it with both hands, he placed its backbone parallel with the saw blade. Carefully, he slid it very briefly lengthwise, backbone against the blade, first on one side and then, turning it over, on the other. He picked up a filet knife, and, laying the fish on the table, explained to the group around him how to slide the knife into the slit made by the saw, run it down along the backbone and out on one side, splitting the flesh neatly off the rib cage. Turning the fish over, he repeated the motion on the other side. He then took out the nugget—the flesh by the fish's stomach on its underside. He held up the final result: two neat, cleanly cut filets of catfish.

"We need to get that cut as close as we can, because that filet's our bread and butter," he told the assembled group. "That's what's going to make the money." He laid the filets in a pan and proceeded to demonstrate again with a second fish. "Now, we need to get those onto a tray and into the freezer as soon as we can."

Glover had been thinking for some time that consumers were ready for value-added products. The first year the company tried it, hardly more than one-half of 1 percent of its total production was sold as filets. Even Farm Fresh's customers did not rush to buy them. Farm Fresh's competitors thought it was an unnecessary move. "My competitors laughed at me," admits Glover, "but it wasn't but a couple of years until they started doing it too." Gradually, the fish-consuming public changed its mind. Today, 70 percent to 80 percent of the industry's total production is in filets.

Catfish were still not widely accepted as a desirable food source in the eyes of the general public, so Glover broadened his marketing efforts. As early as the first or second year of Farm Fresh, he and his wife, Doris, began going to grocery stores and restaurants to cook and give out free samples of catfish. At Piggly Wiggly, Bruno's, and Winn-Dixie, they set up little cooking and sampling stations. Sometimes Joe Jr. also went along.

The Glovers directed their efforts outside Alabama as well as in the state. They cooked and offered pieces of fish on toothpicks to passersby for three days at a World Food Show in New York in the early 1970s. Almost everyone who stopped to look turned up their noses and walked away when they saw that the food being offered was catfish. Even though they had come to an event organized specifically to introduce all kinds of different foods from many different countries, most refused to taste it. "We recognized that it was going to be a slow process," he recalls, "but it had a great deal of potential. It was a quality product, and when people ate it, they knew it was a good fish." In spite of cool reactions, Glover continued to pursue ways to present his catfish to a wider public.

9
More Farmers

> Catfish farming—a truly home-grown industry that is described as American as the Fourth of July and the backyard barbecue—has undergone a vast transition. . . . In those early years, the idea of producing catfish as a farm crop captured the fancy of thousands.
>
> —*Commercial Fish Farmer*

The purchase of STRAL by Nebraska Consolidated Mills, a business with nationwide markets and financial resources, signaled a vote of confidence in the potential of the catfish industry. When Joe Glover established Farm Fresh, he created a competitive market for the fish in West Alabama. The fact that the processors provided harvesting crews made growing catfish even more attractive. The fish farming fever spread, and more and more Alabama farmers took the leap into production.

People followed events taking place around the ponds of the Southeast closely during the late 1960s and early 1970s, but few observers were as yet examining the industry regularly or systematically. Data on pond acreages and production levels were collected by various groups and differ, some of them widely. One estimate suggests that, in 1968, approximately 500 ponds in Alabama were used solely for growing catfish. Two other sources have estimated the number of catfish ponds in the state in 1971 at about 1,440 and 6,000. Both sets of figures appear low in view of a survey of nine southeast Alabama counties conducted in 1969. That survey indicated that roughly 1,100 acres of ponds were used solely for catfish production in that area alone, and those counties were not even in the most heavily producing region of the state—West Alabama. One thing was certain: the catfish industry was growing quickly.

The fledgling industry was not exempt from the same forces that caused the failure of the buffalo fish and dramatic contraction of the bait minnow economies earlier in the century. With catfish harvests steadily increasing in number and size and processing plants sprouting up here and there, many people wanted to grab a piece of the success they saw others having. The sudden swell in production pulled the immature industry out of balance.

Other essential factors in the economic equation were still missing. As a result, far more fish were produced than could be sold, and many farmers went broke. Some found niches and managed to hold onto them. Their stories reveal some common threads.

Bill Kyser's first experience with catfish farming came when his father, Joe Kyser Sr., started the first large-scale commercial fish farm in Alabama near Greensboro in 1965. His father had mortgaged his entire dairy farm to buy land for the ponds. Initially, the elder Kyser had thought that the catfish industry would help pay for the cattle operation. The family had no idea fish farming would evolve into their primary enterprise. Bill, Joe's younger son, enjoyed watching and helping with the work. Before long, they began sharecropping as well as renting ponds in the surrounding area. They turned more and more of their own dairy pastures into ponds. After Bill finished his fisheries degree at Auburn University in 1973, he returned to help work and manage the farm.

Water was available to the Kysers partly because the lay of the land allowed them to trap rainfall and runoff. In addition, Big Prairie Creek ran across their property. They used electric pumps to fill some ponds with creek water but eventually stopped because they could not keep creek fish, plants, and other organisms out of the ponds. Instead, they scheduled their harvests for the fall and relied on winter rains to fill the ponds back up.

The work often amounted to far more than a forty-hour week. Bill Kyser did not mind the backbreaking work or the hours. "It really didn't bother me as much as it did some other folks—working on the weekends and all," he remembers. "I love being on the farm. I love the outdoors. I enjoyed doing physical things. I can't imagine how I did it, now, but it didn't bother me to put a hundred pounds of feed on my back and strike out across the field. I remember the roads getting real muddy and carrying fifty-pound feed sacks, one on each shoulder, out across the creek bottom. It must have been a half a mile from the highway to the fishpond."

There were also times when Kyser had to be willing to go without getting paid. "You do without," says Kyser's wife, Beverly, describing how they managed. They had to be willing to "just keep it to the bare necessities." "In fish farming, in the early days," explains Bill, "you didn't have a regular paycheck and didn't know for sure that you were going to get a paycheck. The pay has kind of come and gone. Several times there wasn't any money there to bring a paycheck home. The fish business has been cyclical, and in the early days when we started, it was really tough."[1] When Kyser had returned home in 1973 to manage the farm, the ponds spanned 200 acres. By 1998 he was raising fish on 450 acres.

J. V. Friday and his neighbor in Selma, Clinton S. Wilkinson, had a standing agreement to let Friday's beef cattle graze on Wilkinson's pastureland. Wilkinson was acquainted with Richard True and Check Stephens and knew about the success of their fingerling business. Around 1966 he and Friday entered into another sharing agreement: Wilkinson would stock a pond on his property with catfish, and Friday would feed and look after them. They would share whatever came of the enterprise.

Friday was excited by the prospect. It sounded interesting. There did not seem to be much risk involved, since he would be paying only for the feed. All he otherwise needed to supply was a little time and labor, and throwing feed into the pond and watching the fish grow did not sound very difficult or time-consuming. This seemed like it might be a job that would blend well with cattle farming and later, after he retired, ought to require less work and time and be easier on him. "I didn't realize what I was getting into," he says with a laugh.

The cattle business did not grow during those years, but the catfish business did. The two men invested in additional ponds on both their properties in the following years. Each year, Friday added a pond, until nearly one hundred of his acres were under water. Gradually, catfish work took more and more of his time, until he was spending half or more on it. Throughout this time, he did all the work himself with only occasional help, until later, when his son was old enough to work for him. In the early 1980s Friday's business expanded into processing as well as production. Together with his son, he converted a small building on his land into a processing plant, selling fish to local residents.

Lenson Montz first got into the catfish business in 1967. He was working at the time at his father-in-law's cheese plant in Greensboro. When STRAL started processing fish the year before in the back of Joe Glover's grocery store, Montz had been one of the many people who had come to watch. The impression left on him by the scene was indelible. "They had seven hundred pounds of fish laying there on the floor," he recalls, "and it just messed my mind up. I decided right then that was what I wanted to do."

Montz's mother, Murvle, owned some low-lying land that could be dammed and turned into ponds. Montz went home and spoke to her about it. His mother was open to the idea and made him the following proposition: "I got the land, and I'll build a pond. If you buy the fish, feed them, and take care of them, we'll split the profits." And that's what they did. Montz approached the venture with great anticipation. "I reckon in the first place I didn't like my job. I was just a country boy—loved the outside—and I always wanted to farm. I always loved to fish and hunt. We were [going to] try it and see if it worked, and if it didn't, we'd do something else. But this was the closest thing I could find to what I wanted to do."

After two or three days of throwing feed into the pond and not seeing fish come up to eat, Lenson and his wife, Ann, were concerned. Their concern melted away, however, when after a few days the fish began to break the surface to take the feed. The Montzes became very excited. Over the next weeks and months, even Montz's mother went down to the bank of the pond during the day to watch. All Montz's father saw was the risk they were taking. "You're puttin' money in a mud hole," he said. Like others, he was skeptical of this "new-fangled" kind of farming.

Everything went well for two years. Lenson worked hard. He put in twelve hours a day at the cheese plant and, with his wife's help, fed the fish in the afternoon and checked them at night. The next year, Joe Glover agreed to buy the fish, and the Montzes sold him their first crop at twenty-five cents a pound. After two years, Montz proposed to his mother that he build additional ponds. She could feed her fish, and he would start his own business. She agreed. Lenson Montz was in the catfish business for himself.

As with many other catfish farmers in Alabama, caring for the fish became a family experience. The Montz children thought the pond was a playground. They enjoyed the water and the fishing. Almost every afternoon, when their father went out on the boat to spread the feed, they accompanied him to the bank, just to be together. Later, when they were old enough to ride in the boat, they would hang their feet over the sides and splash the water. "I fussed at them real bad,"" remembers Montz. "They were making too much noise. The fish wouldn't eat." During the spawning season, the children would help collect the egg masses from the ponds and bring them to the hatchery to be weighed. During the growing season, the family made sure to sample the fish regularly by having fried catfish for an evening meal. Delicious though it was, they made sure not to eat too many. "We wanted to make *some* profit," smiles Ann.

David Pearce's family got into the catfish business in a very small way at first, almost unintentionally. After marrying Fran Weissinger in 1970, he moved the next year with her to her father's row-crop and cattle farm in

Browns, Alabama. Unlike most of the successful catfish farmers, Pearce had no background in agriculture. His college degree was in business. However, since he had not yet established himself in a profession, he was willing to try to learn something new. His father-in-law, Bill Weissinger, was a veterinarian and cattle farmer. Pearce agreed to enter the dairy business and learn about cattle.

Like many catfish producers who began with cattle farms, Weissinger had originally intended his ponds as watering holes for his beef cattle. The bass and bream he stocked in them were just for recreational fishing. He had opened a few of his ponds to the public as a small fee-fishing operation and in 1968 had converted some of the cattle ponds to catfish but had had no serious commercial interest in them.

Along with the dairy farm, Pearce took over management of the ponds, which covered a total of about 40 acres. The catfish side of the enterprise intrigued him much more than the cattle did. Within a year or two, after his first sale to Farm Fresh Catfish Company, he began to focus his attention almost entirely on raising fish. Each year, he added another pond or two, until by 1981 the farm contained about 250 acres of ponds.

Although Pearce had two men working for him, those early years of fish farming made for long, hard hours. He took care of farm management matters, and Fran helped with the bookkeeping and clerical work. With running the farm and raising the children, there was little time to do anything else. "Fran and I went to funerals and weddings, and that was about it," he recalls of their social life.

Pearce's two sons, David and Will, participated in the farm activities. While fish were being harvesting, they enjoyed being out by the ponds. When they were young, they swam more than they actually helped, Pearce recalls, but as they grew up, they worked in every phase of the operation, from harvesting to checking oxygen levels at night. Later they became more involved in feeding and actual management. After graduating from Auburn University, they joined the family business. Pearce's catfish farm became one of the largest operations in Alabama.

Thed Spree's family settled in West Alabama in the 1800s. He had worked on his parents' farm in Boligee since he was a boy. They had raised dairy cattle and soybeans, but in the 1960s, neither of those farming enterprises was doing very well. He and his mother talked matters over. They wanted to find a business in which they could do the bulk of the work themselves. They already had the land, and catfish farming appeared fairly straightforward. Just feed the fish and take care of them, then sell them to the Greensboro plant when they were grown, they thought. Joe Glover had told the Sprees that his crew would harvest the fish, so it looked like an easy way to start a new business. They decided to build two ponds. Spree's mother would keep the books and get 10 percent of what the business made. Spree would do the rest and get 90 percent. That was the agreement.

Spree did not have the money to start the enterprise and had to borrow from the Production Credit Association. In 1972 he and his mother began with a large twenty-acre pond and a smaller one of between six and seven acres. Borrowing for a catfish farm was then, and still tends to be today, an ongoing arrangement that involves refinancing every year. Each time they took another loan and built another pond, they worked hard to be able to pay the loan back completely. Every year, they put back into the business whatever they had made. After about ten years, Thed Spree was, for the most part, finally able to finance his own operation.

The matter of collateral for catfish loans was not a direct or simple matter. Spree and his mother knew they could lose their land if they did not pay back a loan. Everything the Sprees owned was potential collateral and at risk, not just their land. Even the chairs in his house were marked as collateral, says Spree. "If you had a billy goat, they'd put down 'Goat—ten dollars.' They inventoried everything you had. The agent was a good guy. If I told him I'd bought a herd of cows, he'd go out and look at the cows. I told him one time I had twenty thousand bales of hay. He said, 'Worth one dollar apiece?' and I said, 'Yes, sir!'"

The Sprees had difficulty finding employees to help in the ponds. In those early days, people in the surrounding communities were reluctant to work on a catfish farm, especially checking oxygen throughout the night. When asked, people would say, "I'm not going to sit up all night babysitting catfish!" "I can't swim!" or "I'm scared of snakes!" Now things are different, but at the time, people did not have what Spree calls a "catfish mentality." "They didn't want to learn. It takes a certain type of person to get out there. That water is blistering cold. You get out there in waders and work those fish all day, with the wind coming off the water and being that cold. Getting up at one o'clock and working with the seine. Getting out there and fighting the snakes, fighting the ice. I wouldn't want to go through it again."

"The catfish . . . is a repulsive-looking creature, a spiny, bewhiskered bottom scavenger that will eat nearly anything and thrives in some of the most polluted U.S. rivers. Northern fishermen usually throw catfish away in disgust, but tens of thousands of Americans consider its sweet white flesh a delicacy."[2]

These leading words caught Dan Butterfield's eye while he was reading the May sixteenth issue of *Time* magazine in 1969. For a fish lover with an education in biology, the lure was hard to resist. He read on: "especially so when it comes from catfish raised in the comparatively clean waters of a commercial pond." The article was filled with intriguing phrases: "flooding their acreage and raising fish"; "sound financial sense"; "a fatter price per pound than beef, pork, or poultry"; "more profitable than rice, soybeans, and cotton"; "can sell all he produces"; "rising U.S. demand for fish products"; "new products, services, and jobs"; and "thriving catfish industry."

These words were all Butterfield needed to decide that his future could lie in catfish farming. He climbed into his Volkswagen and drove from North Carolina to Texas and then to Arkansas to talk to Edgar Farmer, who was featured in the article. Butterfield was intrigued by what he heard and saw. By October, with money from his parents, he had bought seventy acres of dairy pasture, with a well, near the Black Warrior River. The land's proximity to an urban center, Tuscaloosa, promised a potentially good market for fish. Catfish farming was not yet well established in that area when Butterfield started out. By fall of 1971 his first crop was ready for harvest.

Butterfield is a successful catfish farmer who found an unusual niche for himself, in an unusual way. Unlike most other successful fish farmers in Alabama, fish farming was Butterfield's first true farming experience. He came from outside Alabama, without preestablished roots in any local community. His graduate studies in biology provided him with a good understanding of fish culture and pond management. Like other catfish farmers, he worked hard establishing his business. Though others generally did much of the work themselves, not all participated actively in every single phase of farming, harvesting, and processing. Butterfield did and still does. With the help of occasional employees, he has maintained and seined his ponds himself. He has sold his own fish ever since his second year of operation.

Within a year or two of Butterfield's start with catfish, a rush of farmers entered the fish farming business. Most of them risked and lost. "There was a great euphoria in Tuscaloosa County," he remembers. "It was kind of the only game in town. Everybody wanted to raise them. Virtually all of those businesses failed. Then there was another stampede in 1979. Every time everybody gets excited about catfish, the price collapses. Then everybody gets out." Out of the many in his area who began in 1970, he is the only one who built a successful operation that continued to provide a good living.

In the mid-1960s Lester Taylor and his son, Milton, had intended to harvest the fish in the pond on their row crop farm in Town Creek for the family's use. Once they found that people in the community would buy their catfish, the Taylors learned how to dress the fish in order to sell their first year's surplus. After fee-fishing turned out to be successful for them, they expanded their catfish activities, making it a larger part of their farming enterprise. In 1970 they decided to build a fish market.

Northwest Alabama's hills provided plenty of water to fill ponds, but the Taylors did not have enough sites suitable for hill ponds. Their operation grew more by renting and leasing ponds built by other landowners in their area. Finding demand growing steadily over the years, they looked around and found catfish to buy from other pond owners.

Catfish farming has continued in northwest Alabama on a relatively small scale. The Taylors' farm is special in that it is associated with their small but thriving processing business and market. Also somewhat unusual for the contemporary industry, they still harvest and process the fish with their own crew. Although different in some ways, their business shows characteristics typical of the farms that were becoming established elsewhere in the state. Initially, Lester Taylor borrowed money to get started. For a number of years after getting into the business, the Taylors continued to operate their fish business as just one part of their total farm operation before turning exclusively to catfish. They drew on harvests from rented and leased ponds as well as production on family farmland. It was a family operation, with Milton Taylor sharing the work with his father and his wife, and at times with his mother and other family members.

These are only a few of the Alabama farmers who have operated successful catfish farms over the years. In some ways, their initial circumstances and later experiences were very different from one another. Yet there are also certain common patterns.

Those who took the leap into catfish had to be willing and able to take on financial risks. No farming, and *especially* not fish farming, is ever risk-free. Few had a great deal of capital to begin with. "Those who had made money were those who had almost nothing with which to begin," according to comments in the *American Fish Farmer* in 1970. "Their mistakes were small ones from which they learned. Those who got into the business with plenty of capital began by making a series of big mistakes and were soon out of business." They needed to enjoy the out-of-doors, and they had to be willing to learn new things. In particular, they needed to learn from their experiences. "The people who have been successful know there are few text-book answers, that every inch of earth and every drop of water is different," maintained the author. "Even ponds, side by side, react differently and fish production will vary in the same pond from year to year."[3]

Always, those who succeeded worked very hard. All had access to land—land that would hold water. Some of that land was their own or family property, and some they rented or leased. Typically, kinship relationships were central to obtaining access to or ownership of the land. Participation of entire families was common. Often, kin provided initial financing. Nearly all the farmers were intending, at first, to add a new element into already existing farming operations, and fish production eventually became their sole activity. Soon, almost all of them not only had food fish in grow-out ponds but also their own hatcheries.

Notes

1. William T. Kyser and Beverly Kyser, interview by author, tape recording, 19 January 1998, Special Collections and Archives, Auburn University, Auburn, Ala.

2. "Food," *Time*, 16 May 1969, 112.

3. Roy Prewitt, "Rambling Along," *American Fish Farmer* 1 (March 1970): 23–24.

10

A Hatchery for Everyone

> It is the same old story we've had throughout the industry the past few years. Once the technique of producing fingerlings was vastly improved, it made it very easy for every Tom, Dick, and Harry to grow their own. They had been paying good prices for fingerlings, so they felt they could produce their own and sell them at a good price.
>
> —Richard True

The supply of fingerlings fluctuated during the early 1970s with the generally unstable catfish market. Some years, fingerlings were available in large supplies at low prices; in others, the supply was short and prices were up. Commercial hatcheries did not always have enough fingerlings of the size farmers wanted, when they needed them. Delivery expenses were added into the cost. Fingerling production looked more and more attractive to many growout farmers.

Other forces worked to stimulate widespread fingerling production also. An air of optimism about the future of the industry prevailed. Establishment of processing plants in West Alabama expanded the market for food-size fish. Many farmers were building additional ponds. Guidance was available from fisheries biologists with the state Division of Game and Fish, the SCS, and individuals like O. L. Green, a biologist who had worked at the Marion hatchery and was now employed at Farm Fresh. An increased understanding of pond and fisheries biology had been developing as a result of work being done by Auburn University's fisheries researchers, and that expertise was being mobilized in order to make it available to farmers. Professors and staff began paying visits to pond banks. They began teaching short courses on the ins and outs of fish farming, including spawning, hatching, and rearing of fry.

Many farmers had a small building or structure on their land in which they could build a hatchery, a place where they could control conditions adequately and watch over eggs and fry. If not, they built one. Often, hatcheries were little more than a roof supported by partial walls. Electric paddle systems aerated the spawns in the troughs. The fry were transferred to ponds to grow to stocking size.

A lot of do-it-yourself invention and experimentation took place. Growers observed or learned of spawning methods used by others and adapted them to their own situations. Both research and experience with generations of brood fish had shown that, given proper conditions and containers in which to lay eggs, catfish would reproduce adequately on their own in the still water of ponds, so many placed milk cans, such as those used by STRAL, in spawning ponds. When milk cans were not available, they tried other containers. Buddy Lavender managed hatcheries on more than one farm over many years. He first adapted 30-×-16-×-9-inch ammunition boxes for spawning use. "I'd take the last fifteen inches and burn it off," he explains, "punched holes in it, and tied wire in there. . . . In the early spring when the water is cool, you just barely cover the tops of them. You can just flip it up and reach in there and get your pad of eggs out without having to pick them up. I've tried wooden boxes, and now we've got plastic jugs that we get from the soap factories. You try everything. It depends on what you can get."

Bill Easterling's first spawning containers were built by local high school students, members of the school's Future Farmers of America club. The boxes were modeled after what he had seen used at Auburn University's experimental ponds. The students "built spawning boxes for us out of rough pine lumber," he remembers. "They were probably eighteen inches wide and around twenty-four or twenty-six inches long. We had a lid with hinges on them. We turned back the lid, felt down, and felt eggs. We had been told how the eggs should be taken from a container, and we were real quick to learn how to do that effectively."

At the very minimum, a farmer could "wild spawn." He needed only brood fish, containers, a pond, and possibly a pond to which to return the brood fish later. J. V. Friday spawned his catfish fingerlings in this way, requiring a minimum of expense and time and making it unnecessary for him to mortgage his house in order to finance a larger, more complicated hatchery. The fingerlings he raised were primarily for his own use. Every year, Friday chose his broodstock and put them in a small pond. "It had to be real clean from any other fish in there," he explains. "We put containers in there in May, June. The fish would lay the eggs in it. They'd hatch right in those containers. A lot of times we just left the brood fish and the fry in there until they got larger. And then sometimes we'd go and take the brood fish out and leave just the fry and grow them up until they got large."

The main disadvantage to wild spawning in the open pond was that it was almost impossible to tell how successful the spawning and hatching had been. Another disadvantage was the danger of predation. When water was pumped from creeks or other surface sources, other types of fish could enter the pond and eat the very young fish.

Harry and Neal Miller began raising fingerlings themselves because they needed more for their own use. Like many others in West Alabama, they received advice from O. L. Green at Farm Fresh. Under Green's guidance the Miller brothers began placing large cans in a pond. The brood fish they had selected laid their eggs in the cans, just like Green said they would. Every other day or so during spawning season, in late spring and early summer, the brothers left their growout fishponds, honey, and other farming activities to go to the spawning pond, lift the cans up, and go through them for eggs. Rather than remove the brood fish with a large-mesh seine and leave the fingerlings to grow in the pond, the Millers preferred to remove the spawns and take them to Richard True in Newbern. The brothers felt better having him watch over the hatching process. Eventually, the Millers built a hatchery on their own property and were able to gain a little added income by selling their surplus fingerlings to other farmers.

Lenson Montz also built a hatchery for his own use. Rather than buy several hundred thousand fish every year for his seventy-acre operation, he thought he might be able to raise the fingerlings more cheaply himself. The Montzes were among the many fish farmers who set up hatchery operations with the help of SCS biologist David Kelly, who traveled every week from Auburn to Greensboro to work with them. They raised fingerlings for the rest of their fish-farming career.

For farmers with large properties and an adequate source of water, starting up their own hatchery was easier than for those with more limited land and water resources. For all of them, however, an essential element in their operations was the participation and support of family members.

When Bill Kyser graduated from Auburn University in 1973, he knew he wanted to return to Greensboro to be a partner in his father's business. The nearby STRAL hatchery in Newbern had been established twelve years earlier, but True, working now for ConAgra, had recently gone to Mississippi. The hatchery still belonged to ConAgra but had been closed down. The Kysers rented the ponds and the building, bought broodstock from ConAgra, and started their own joint venture hatchery.

Bill ran the hatchery with his father's help and lived with his new wife, Beverly, in True's house while True was away. Beverly taught school and supported the two of them. Even though Kyser had a college degree in fisheries, running his own hatchery proved to be a challenge. "Hatcheries had come and gone," he recalls. "There was still a lot that nobody knew. . . . The first year we thought we had really made it, because we had hatched off a million and a half fingerlings. We raised them to three or four inches with very little problem. Then, in the early fall, about three-fourths of them died . . . just plain disappeared. Ich and a few more diseases hit us that fall that we really didn't know about . . . weren't prepared for. We never even knew we didn't have [the fish] until we started seining the ponds."[1]

When the lease ran out in 1976, ConAgra offered to sell the Kysers the hatchery. They had not made any money and had no way to pay for it or to make it pay for itself, so they decided not to buy it. After three years working the hatchery and Joe Kyser's food fish farm, Bill and Beverly built their own small hatchery on his father's property and set out on their own.

A significant proportion of farmers' wives became more involved in the catfish business than they had been before. Hatchery work seemed well suited for women. They may have previously helped with record-keeping and other aspects of farm paperwork, but the hatcheries now provided a niche for those who enjoyed working directly with the fish. The scale of the work was physically more manageable and attractive for the women than the heavy labor at the ponds. Many farmers feel that their hatcheries benefited from the care the women took with details and from their training and experience with keeping things clean. It was crucial that the troughs and equipment be kept sanitary if spawns and fry were to stay healthy.

Since it freed the men to concentrate on other aspects of the farm, the division of labor seemed to work well. "Once we started our own hatchery," says Beverly Kyser, "I just started doing the work [there] . . . making sure everything was running. If there was a problem I'd call Bill, but Bill tended to [the fish] when they went to the pond. Just keeping day-to-day tabs, treating the eggs, siphoning the fry, feeding them. I'd do the in-hatchery work, and he would do the out-of-hatchery work."

Hatcheries provided an opportunity for women to apply organizational skills similar to those they used managing their households. After children were old enough and needed less frequent attention, some women effectively took over and managed the entire operation. "My wife pretty much runs the hatchery herself," says Randy Hollingsworth, who is of the opinion that women generally do a better job in the hatchery than men. "Her efforts will be completely concentrated on the hatchery. When I was trying to run the hatchery and see about the farm also, I would have things that I said were pushing harder than that hatchery work, and I might not do as good a job. Whereas, when I turned that over to Emily, she's going to see that that job is completed that day, and ready for the next."

Many of the women found gratification in caring for the young fish. For women who took care of children and households, working in the hatchery offered a change of pace. Though it was not her only job, Ann Montz enjoyed working in her family's hatchery. "It was very interesting," she says. "I would draw the little bitty fry off and clean it up. I loved to clean it up. The cement floor would get fish and fish eggs, and it would be my job to clean them up with a power washer."

Somewhat the exception to most fish farmers' wives, Ann did not confine herself to working just with the fry. Her husband, Lenson, testifies proudly to the pleasure his wife took in other aspects of the hatchery operation: "She'd go out in them ponds and gather eggs, too. They put ten-gallon milk cans out in the ponds for the fish to go in and lay, and she'd go out and run the ol' fish out and steal the eggs away from them. A lot of times, she'd take her hands up in the can to get the eggs, and the old fish would bite her on the hands. She'd scream and holler, but she'd run them out and steal the eggs anyway."

Catfish farming in Alabama was a family operation in yet another way. The hatcheries were busy places during only a few months of the year, in the late spring and early summer. After school, on weekends, and during summer vacations, children often worked along with their mothers. "They loved it," remembers Ann Montz of her children. "We all loved the hatchery. They would go get the eggs out, when they got older, and bring them up here. They'd help us." During hatching season in particular, the help of the children could be essential to the farm's success.

~

Late one evening in May, around 1975, on the Montz farm:

The children were finally asleep in their beds; the kitchen had been straightened up for the next day; and the farm record books had been brought a little more up to date. Rain from an earlier thunderstorm still drummed on the roof. Lenson Montz finished watching the late news on television while Ann checked on the children and then went over the next day's plans in her mind. After the news, they got ready for bed. Ann pulled the sheet up around herself with a sigh, while her husband leaned over to turn out the light and then joined her for a good night's rest after an exhausting day out at the ponds. Away from the house, in the hatchery, the paddle wheels dipped and turned, pushing the warm water slowly past the baskets of yellow and pink spawns.

Suddenly, in the middle of a deep sleep, they were awakened by a siren. Lenson's brother Dathan had put together a battery-operated alarm system to warn them of power outages. Whenever the electricity went off, mercury in the switch ran down and set off the siren. Still half asleep, Lenson jumped up out of bed, pulled on his blue jeans and shoes, and ran out the door into the darkness. Ann jumped up as well, threw on a housedress, and ran to get the children up.

"Bubba, wake up!" she called softly as she gently but urgently shook her son, Aubrey Jay. As soon as he began to stir in his bed, she turned to her daughter. "Get up, Anna! The electricity's out! We need to help your daddy!" she said, always amazed that they could sleep through the alarm.

After they all had shoes on, they ran out to the hatchery. The Montzes knew too well that the millions of eggs and tiny fry in the troughs could all die if they did not get oxygen within ten minutes. Nothing needed to be said. Some years, spawning season was punctuated weekly by emergency trips like this to the hatchery, particularly during stormy weather. Hearts pounding, both Ann and the children knew what to do. They spread out among the troughs and started cranking paddle wheels by hand. The water had to be in constant motion. The children leaned into their task, pulling around and around to turn the wheels, until their arms got tired and they were allowed a brief rest. Now and then, everyone moved to a different place to keep the water flowing in all the troughs. Talking encouragingly through the noise of the rain and wind, Ann watched that none of the baskets got accidentally dislodged.

The darkness all around indicated that the outage was affecting more than just their property. Occasionally, one of the wires powering the paddle wheel motors might have been blown loose in the storm and caused a short. Lenson Montz checked for this possibility, but as was usually the case, the problem tonight was not one he could fix himself. "Keep on turnin', children," encouraged Ann. "I know you're tired, but we have to keep paddlin' till the electricity comes back on."

The Montzes had no generator for a backup power source. Generators were too expensive an item for their budget. Lenson joined his family at the troughs and began turning paddle wheels. He hoped to stock the fingerlings in his growout ponds later that year, and he could not afford to lose them.

They stood there a long time that night, getting back to bed only hours later, after the storm damage had been repaired and the electricity was back on. Today, having a generator is a relatively cheap form of insurance for hatchery operators.

Caring for the hatchery required round-the-clock vigilance. The electricity went out not only at night or during storms but also in the daytime, even in fair weather. Farmers' wives who managed hatcheries had to respond immediately, no matter what they were doing at the time. Wherever she was, when Beverly Kyser would notice lights and motors going off, she would take off running to the hatchery, which was conveniently near their house.

The Kysers got used to sleeping with just the screen door latched, so they could hear any changes in the sound of the motors during the night. Their ears had become attuned to the little sounds coming from the paddle wheels. Two or three times a night, Beverly would get up, throw on her bathrobe, and walk outside the fifty feet or so to check on the system, making sure everything was all right. It was a lot like having a newborn baby. Sometimes, she awoke to hear nothing but the rasping of the katydids and the crickets. She would jump out of bed and go to the troughs to find that a basket had shifted and gotten caught in the paddle wheels.

In those early years, like many of Alabama's fish farming families, the Kysers often had their young children tagging along behind them. When they walked through the hatchery, the children tended to run their hands through the water and slide baskets around. The children were not the only ones who could cause problems with the equipment, however. Interested visitors would come over to look, and if one of the metal baskets had a rough edge, it might catch someone's clothing and get pulled aside instead of staying straight and centered.

Beverly Kyser simply could not leave the hatchery for very long. She had to be aware, keeping tabs on everything, because sooner or later something would happen. "With the eggs it doesn't take very long—ten, fifteen, twenty minutes, depending on how old the eggs are," she explains. "The real ripe eggs will fall apart a whole lot quicker than the younger ones, so you just have to be there."

Activity at the hatchery went in waves. Warmer temperatures prompted earlier laying and, up to a point, hatching. By the second or third week, the numbers of spawns would increase, and they continued increasing until they peaked around early June, depending on the weather. If temperatures got too high, laying would stop. One never knew from one day to the next what would happen with the weather. Things have not changed. One still has to be constantly on watch. "You hope you've gotten everything up to that point that you need. You don't leave town. Somebody is [always] there," according to Beverly Kyser.

The time needed to produce a crop of fingerlings may not seem very long, but it is a period of intense activity. The spawn, fry, and small fingerlings go through many more changes during that time than do more mature fish. They are very vulnerable to organisms that may have little effect on larger fish. For months before and after spawning and hatching, there is no time when conscientious fingerling producers are without worries, keeping on the lookout for many different potential problems. "You start out with a brood fish that's got to be healthy," according to Bill Easterling. "It's got to be properly fed so it will be productive. . . . You've got to bring them in. You got to take eggs. You got to know what water temperature those eggs will take. You got to be able to run lab checks on those fingerlings until they're ready to be marketed. And when you market a fish, you got to have sufficient knowledge to know what you're selling and that you're selling a healthy product. A fingerling producer has to be real sensitive to [predatory] birds . . . from the time he hatches until he's marketed them."

Disease and parasites were a more constant threat than electrical outages to hatchery operations. A fingerling producer could find himself needing expert advice every day during the spawning and hatching season, but he could not leave to look for it, according to Easterling. "Putting a lot of small fish, crowding them into an area, lends itself to an outbreak of parasite or bacterial infection. It's not a day-to-day problem with the pond owner as much as it is with a fingerling producer."

For producers dependent on income from fingerling sales, basic knowledge about the organisms threatening the young fish was essential. Many fingerling farmers learned from short courses, from researchers, and from extension agents how to take better care of their fish. They learned to do things like conduct basic tests for water quality and for parasites.

Keeping a hatchery clean of disease agents was not then, and is not today, an easy thing to do, according to Randy Hollingsworth. Sometimes, using well water eliminated parasite problems, but well water was typically too cold for the egg masses and fry. Most often, the organisms came from using existing surface waters. Visitor traffic through the hatchery brought risks also. "It's so easy to spread disease," says Hollingsworth. "There are so many fish in such a confined area. You don't want to take a chance of somebody putting their hand in a tank and trying to pick up a fish and look at it and maybe injuring it or transferring a disease from one tank to another."

Producers operating hatcheries faced many issues and problems also faced by growout operators. The successful ones looked actively for information and assistance. Hatchery operators who knew how to forge associations with fisheries specialists and with other fish farmers benefited directly from the sharing of knowledge and experience. Farmers were constantly learning from each other.

Cooperation and sharing of information also resulted in benefits such as trades of equipment. Easterling obtained a paddle system for his hatching troughs by offering fingerlings in trade. "There's always interchange of information and equipment," he says, "each individual looking at what the other was doing [effectively] with a piece of equipment. . . . Then you would copy it. Ninety-nine percent plus had always been very free with information they had. We kind of grew up with everybody looking for information, willing to share what they had."

Around Greensboro, people knew *of* each other even if they were not directly acquainted, and a real sense of cooperation existed after the industry caught on, according to Bill Kyser. "Back then, it was much more difficult than it is now to hatch [catfish], because they were working basically with wild strains. Now we have domesticated strains. In general, most of us did not consider each other competitors as much as we considered ourselves competing against wild-caught fish, poultry, and hogs."

Farmers raising more fingerlings than they could use themselves had to find a market for the extra fish. Some of the young fish went to private pond owners who grew them for commercial purposes or for their own family's use. Some went to lakes and other bodies of water used for sportfishing. Almost everyone raising catfish in Alabama knew others doing the same, so there was little need for advertising.

The distance over which a catfish farmer could transport his extra fingerlings for delivery was limited, but by the late 1970s at least, some hatcheries began to sell internationally. Bill Easterling delivered for several years as far away as Central America and Europe, having found customers in Guatemala who wanted two-inch fingerlings and a market for larger ones in Italy. However, regulatory and legal requirements of shipping the fish abroad eventually became so constraining that international transactions were not a significant portion of his, or the industry's, sales.

Many farmers who managed to raise healthy fingerlings found the enterprise reasonably lucrative. Some eventually left the business because of disease problems. To most, the business was for many years like the growout part of the operation: hand-to-mouth. New loans were taken out every year to continue or to expand operations. Money made in sales typically went back into the hatchery or into the farm. Eventually, operating both a hatchery and a fish farm proved more effort than it was worth to some farmers. Selling fingerlings may have provided some additional income, but it also brought additional work. Some farmers who had expanded into hatcheries continued, but not commercially. As Randy Hollingsworth puts it, "We're in the fingerling business now just for ourselves. We sell a few, but it's very few, because it's not a good money-making thing." Many small and midsize growout operators came to the same conclusion.

Note

1. The protozoan parasite Ich (pronounced "ick") can cause high mortality rates in infected ponds. Caused by the organism *Ichthyophthirius multifiliis* (hence the name "Ich," for obvious reasons), it occurs most often in cool weather during the spring and fall.

11

Rising Production, Rising Problems

> In the catfish business, you don't get second chances, especially when you've got oxygen problems. You either meet the demand right then or that's it. You've got to perform. It just about wore me out. I'm like an old truck, you know. I'm fifty, but I've got a lot of miles on my speedometer, because it's pretty strenuous. You get out there, and you don't know if you're going to lose those fish. You're doing all that you can do, and you got a hundred thousand pounds of fish out there, all coming up.
>
> —Thed Spree

Methods and technologies of production underwent significant changes during the industry's first two decades. One such area was pond design and construction. Nearly all the ponds built in Alabama had been intended originally for watering cattle or for recreational fishing, not for producing a commercially viable product. The first catfish ponds had been created with nothing more than earthen dams between two hillsides. Most of these early ponds were on land that couldn't be farmed—often lowland or wetland areas that served well as catch basins for runoff water. Requirements were only that the sites held water and crossed no property lines.

Only by experience did people learn that those areas were, according to Elvin Wright, some of the "poorest places you could build a pond for catfish." The ponds were problematic not only for harvesting but also for growing the fish. "Some were too shallow," remembers Wright. "In the hot summertime, the water would heat up a little too fast. You were fighting weeds all the time because you didn't have enough depth."

Gradually, learning from the problems encountered by farmers and harvesters, engineers developed designs that functioned better for commercial aquaculture. New ponds were built more uniformly deep, with cleared-out bottoms, so that they would not need to be drained down for harvesting. By the mid-1970s, when production had increased to the point at which the plants often had plenty of fish, processors began to refuse to harvest the older types of ponds. "You need to fix those ponds," they told the farmers. "We aren't going to harvest them any more until you clean them up."

As production increased and harvesting became more mechanized, reliable access for large vehicles and equipment to get near the water became important for both farmers and harvesters. Previously, Wright explains, "if the weather was bad and you couldn't get to the pond, that was all right. But with a catfish pond, you have to be on that pond every day, regardless of weather, just about, and you've got to be on the pond with heavy equipment. So you want an area that will hold your equipment at any time." Where ponds could be placed next to each other, expensive road-building was kept to a minimum. Levees were built to permit travel between and around them with heavy feed trucks, harvest trucks, and tractors. Farmers who could afford it began to top the levees with gravel to allow access rain or shine.

~

The design and construction of ponds affected harvesting methods. Early harvests had relied almost entirely on pure manpower. The seines that gathered the fish were difficult to drag through the water and also to lift, wet and muddy, out of the water onto trailers to haul between ponds. "We used to have to pull seines by hand all the way around the pond," remembers J. V. Friday. "The farther you went, the heavier it got." As more of the new ponds were built, longer and wider seines could be used. Eventually, the seine reel was developed. Drawn and powered by a tractor, it was large enough and powerful enough to play out, drag, and reel in very long seines.

Once large ponds could be harvested mechanically without being completely drained, however, more fish escaped under the nets. The escapees stayed in the pond to grow larger than ones stocked later. This staggering of ages caused problems for the processors, who needed to have fish of relatively uniform size. Picking out the size they needed by hand was impractical for the crews.

In 1970 Paul Smith, a graduate of the Auburn University fisheries program who was working for ConAgra in Mississippi, hired a fellow graduate by the name of Wayne McLendon to manage the hatchery operation at ConAgra's sixteen-hundred-acre fish farm in Tupelo. At Auburn, McLendon had had conversations with Homer "Rudy" Schmittou, a graduate student who had been researching the use of cages for raising catfish. One topic in their discussions had been the matter of separating cage-raised fish by size. Those thoughts led later to ideas for separating fish raised in ponds.

A primary concern during harvests was minimizing the stress the fish experienced while being separated and held in the water. McLendon and Smith bounced ideas off each other. They envisioned creating a type of "throat" through which the fish could swim on their own from the seine to a holding net. The two men discussed the problem with experienced harvesters and with seine manufacturers at Delta Net and Twine. Together, they came up with a plan. Stakes long enough to project above the surface were sunk into a pond bottom. Hooks at the top ends provided places to anchor a net. The throat they settled on was a five-foot-by-three-foot rectangle. On the other side of the throat from the seine was a closed net sized to hold only the larger fish for harvest. This net was called a "grading sock."

McLendon returned to Alabama in 1975, but not before he and Smith had established the use of the grading sock as part of the method now called "topping off," which has since become a standard harvesting technique. Thanks to topping off, harvest crews could selectively take uniform sizes for processing; the fish experienced less stress; and farmers were able to restock ponds and have multiple sizes of fish available year-round.

The development of feeds was an extremely important factor in increased production. Feed companies had conducted their own research efforts along with university and government scientists who prepared the way by studying nutritional requirements of catfish. The result was more efficient feeds that promoted healthier and faster-growing fish, better water quality management, and greater economies for farmers' pocketbooks.

Experimentation on fish feeds had begun at Auburn University as early as 1934 and continued into the 1950s under the leadership of Homer Swingle. By the mid-1950s Ellis Prather had developed a practical, well-balanced feed for warmwater fishes. It was dubbed "Auburn No. 1."

When Bob's Feeds in Brundidge, Alabama, acquired a pellet mill in the early 1960s, the fisheries program at Auburn began to have its catfish feed pelleted. Pelleted feeds were more desirable than nonpelleted or meal-type feeds, because less was lost when the feed was put into the water. Prather's pelleted mixture was called "Auburn No. 2." Much higher in protein than the feeds used today, it contained no supplemental vitamins and still left much to be desired, but at the very low stocking rates that were used in the early 1960s, it "worked fairly well," according to Tom Lovell, an Auburn professor who researched fish nutrition and feeds from 1969 to 1998. Later, a third, even better mixture would be formulated.

In the late 1950s Homer Swingle had already recognized the need for guidelines for growing channel catfish and started experiments on catfish production for commercial purposes using Auburn No. 2. His results indicated that by stocking two thousand to three thousand four- to six-inch fingerlings per acre in the late winter or early spring and feeding daily until the following fall, farmers could produce as many as two thousand to three thousand pounds of marketable fish in a one-acre pond. Swingle had offered his findings as guidelines for farmers, with estimates of how many fish could be stocked safely in a pond with a given surface area, what should go into a catfish feed, and how much feed should be put out. Soon, Auburn No. 2 began to be used by Alabama's producers.

Because natural feed ingredients such as soybean meal and corn are not complete in vitamins, Auburn researchers conducted experiments to determine the effects of supplemental vitamins on the fish. Not long after, Bob's Feeds began premixing vitamins into Auburn No. 2. The addition improved fish production significantly.

Over time, researchers zeroed in further on desirable nutrient combinations for the feed. Scientists at Kansas State University had started experiments in the late 1950s and 1960s, but the Kansas climate and funding problems limited what could be accomplished there. Harry Dupree had done his dissertation research at AU on vitamin requirements and continued work later on amino acid requirements of catfish. Other contributions were made by Auburn researchers from the late 1960s into the 1970s and later at laboratories in a variety of other states in the South and the Midwest.

When Tom Lovell came to Auburn in 1969 to research fish nutrition, the channel catfish had become the most important cultured fish in the South and was, therefore, a logical target for feed research and development. Most of the catfish's nutrient requirements were determined in the years immediately after. That work established the guidelines for formulating the practical and efficient feeds and feeding regimes available today. Because about 50 percent, on average, of a catfish farmer's expenses can be attributed to feed, it is commonly agreed that the development of cost-efficient feeds has been a major factor in the success of the industry.[1] No other commercially cultured fish receive a feed as cost-effective and high-performance as channel catfish do.[2]

Cooperation between farmers and commercial feed manufacturers started early, and feed companies began competing around 1964 for the catfish feed market. The STRAL Company offered feeds and feeding instructions to its customers along with its fingerlings, but as a long-time feed salesman,

Check Stephens understood that the Hog Chow they were recommending to their growout customers did not necessarily offer an ideal formula for catfish, and Trout Chow was very expensive.

Western Grain Company in Birmingham promised Stephens it would formulate a feed for the fish, the first such commercial feed produced. The STRAL partners agreed in turn to sell the new Western Grain feed. Almost at the same time, the Specialty Chows department at Purina contacted STRAL. When Stephens suggested "Catfish Chow" as a name for Purina's new product, the name stuck.

The commercial feeds were made in a dense pellet form and always sank when broadcast on the water. When the Illinois feed company Ralph Wells developed an extruded catfish feed called "Splash!" in the late 1960s, production received a huge boost. The extrusion process expanded the pellets so that they floated. Farmers found that their fish ate the extruded feed very readily.

Floating feeds were a significant improvement. They were a great management benefit. Because the fish came to the surface to take the floating pellets, farmers were able to tell exactly how much feed was being consumed. Previously, farmers had been buying and applying more feed than necessary. Floating pellets lessened the amount of overfeeding. Less feed and, therefore, less of the farmers' money was wasted. Excess feed always increased the risk of oxygen depletion in the ponds and was a factor in degraded water quality. The opportunity to observe the fish at the surface also helped farmers in other ways, including watching for disease outbreaks.

Homer Swingle's 1958 guidelines had recommended that farmers limit stocking to two thousand to three thousand fish and give them no more than thirty-five pounds of feed per day per acre of water. These were considered safe quantities for keeping wastes at a level that pond organisms could assimilate without causing oxygen depletion. Most people had been stocking their ponds accordingly, and though serious oxygen depletion did occur occasionally, it was relatively rare. With improved feed quality and feed conversion ratio and the ability to see how much the fish were eating, farmers found that they could stock more fingerlings and grow more fish using the same amount of feed and pond space.[3] The vision of larger harvests and larger sales at the end of the growing season was too tempting for many farmers to resist, and in the mid-1970s, stocking rates in the industry jumped noticeably. Crowding four thousand and even five thousand pounds of fish into an acre of water was not unusual.

Higher stocking rates produced problems, however. More fish meant more dissolved oxygen was needed. The increased fish wastes only added to the demand for oxygen. Before long, oxygen depletion became the primary limiting factor in production and almost proved to be the undoing of the entire industry.

Out on the pond, spreading feed from their boats, farmers were able to see changes in the water quality of their ponds. However, they did not know what those changes meant. They went out each day, and if the fish did not eat, they figured something was probably wrong. If the fish ate fine, they saw no reason to worry. If the day was particularly hot, a farmer might check the pond later, but if he had no reason to suspect any problems, he might do nothing till the next day. "The first year we didn't know hardly anything about the fish business," admits Lenson Montz. "We just learned by mistakes. . . . We didn't know what oxygen trouble was. One day the water started turning real green. The water was real pretty, we thought. The fish was eating good, but one day we went down to check on our fish, and all the fish were [up] trying to get some air. They come to the top, because ripples from the water makes a little bit of oxygen."

As Thed Spree puts it, "We were just plumb dumb. That's all there was to it. You can get away with being dumb for a year or so on a new pond, maybe, but it will catch up with you the next year, or the next." In fact, when oxygen problems caught up with a farmer, they did so very quickly. He might return to a pond an hour or two after having found nothing apparently wrong, only to find twenty thousand pounds of catfish lying dead on the banks and floating on the water (see fig. 10). When that happened, there was nothing to do but pick up the ten tons of smelly, dead fish by hand and think about all the money he had invested in fingerlings and feed that was now completely lost.

> You know, there's a lot of money to be made in catfish, but it's at such a high risk. I mean, you can go to bed rich one day and get up brokest thing in the world the next morning.
>
> —Buddy Lavender

In the early 1970s there were not many places to which catfish farmers in West Alabama could look for advice when their ponds "turned bad." Because of the industry's rapid growth a time lag resulted between the problems farmers experienced and the establishment of research and technology geared to adequately serve their needs. The three main sources of aid came

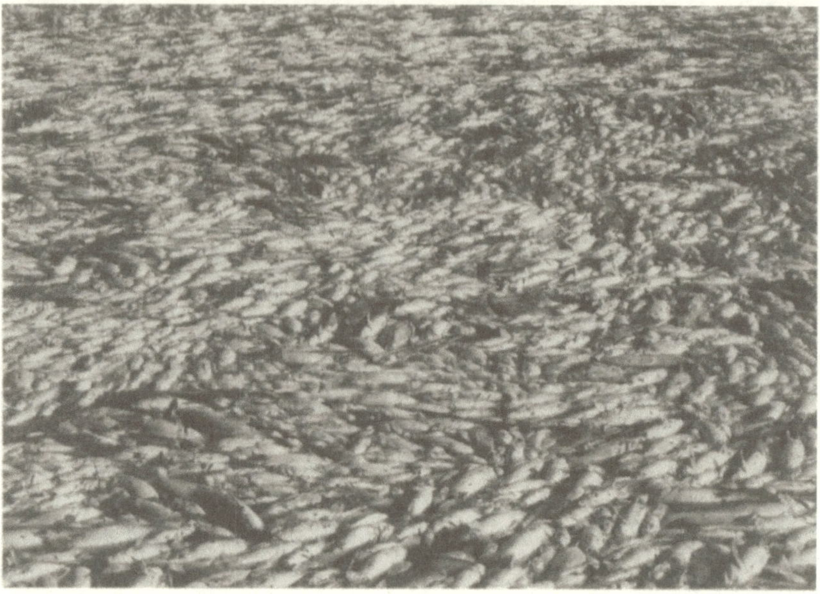

Fig. 10. Massive catfish die-off in western Dallas County, 1998.
The cause of the die-off was insufficient dissolved oxygen. Fifty thousand pounds of fish were lost, at an estimated value to the farmer of thirty thousand dollars. (Photo by: Gregory N. Whitis, Alabama Cooperative Extension System.)

from the FWS fish hatchery in Marion, the SCS, and Farm Fresh Catfish Company. Certain employees there took it upon themselves to go out of their way to help. Their interest and investment of time and energy filled some of the gap.

The biologist O. L. Green had worked as a fish culturist at the hatchery in Marion and given information and encouragement to fish farmers since the beginning of the STRAL Company. Although knowledge about commercial production and catfish health was still limited, he had visited their ponds and shared with them what he had learned from his broader experience in warmwater fisheries. He appreciated the farmers' plights and their aspirations and continued helping them even though his job did not include outreach responsibilities. To the farmers, he was very approachable. "You could talk to O. L.," says J. V. Friday, echoing the sentiment of many early producers. They came to rely on his straightforward attitude and advice.

A catfish farmer himself, Joe Glover recognized the importance of understanding the relationship of oxygen balance and fish health. He was well aware of the difficulty farmers had in finding information and assistance, and he determined to do something about it. When Green was ready to retire from the federal hatchery in the early 1970s, Glover offered him a position as consulting biologist with Farm Fresh in Greensboro. With the office and laboratory Glover provided for him at the plant, Green began immediately to work full-time with the farmers who did business with Farm Fresh, at no cost to them. The free service Glover and Green provided many West Alabama farmers was an important support for the industry at that time. "People would take their fish in to O. L., and he would check the fish and the water quality," recalls Buddy Lavender. Green performed informally the same service that fisheries extension specialists would be doing later on an institutionalized basis.

For years, Green gave the farmers an education as well as advice, teaching them informally about the cycle of events that gave rise to oxygen-deprived ponds. He taught the farmers about parasites and bacteria, infections, and all kinds of diseases the fish would have. Many viewed his assistance as a lifesaver. "He would visit the farmers, give them advice on stocking rates, how to treat ponds, how to deal with off-flavor," recalls Joe Glover Jr. of Green, who died in 1998. "The farmers believed in him. They really liked him. He was plainspoken. He told you what he thought, but he was really respected. My father didn't have to hire him to be a good PR guy. Farm Fresh didn't really need a PR guy, because it was the only plant here for several years. The farmers miss him."

A second individual whose efforts helped significantly to fill the gap in outreach was Dave Kelly, a graduate of the bachelor's and master's programs in fisheries management at Auburn University. Kelly took a job with the SCS in 1973 and in 1978 became what was then the only SCS fisheries biologist in the nation. Supported by his immediate supervisors, Earl Norton and later Ernest Todd, in a position that had no history or tradition, he had few restrictions placed on him. Kelly had freedom to put his energies into whatever he saw fit. Working out of an office in Auburn, he chose to travel regularly to Greensboro and stay a week at a time, helping area landowners and fish farmers manage water quality in their ponds. He always tried to "be there when they hollered."

Kelly's conscientious work was an important element in the progress of Alabama's catfish farmers for more than two decades. During the years after farmers started increasing their stocking rates, when the oxygen balance in their ponds became tenuous during the hot season, he could be found on many a sultry summer night with a farmer on the dark banks of his ponds, talking about the situation and helping the farmer through to the next morning. His interest, encouragement, and "hand-holding" helped many West Alabama farmers later handle similar situations better on their own. "They got a lot of confidence just having him around," in the estimation of Wayne Shell, then head of AU's Department of Fisheries and Allied Aquacultures. "He would spend an awful lot of nights out there, all night long, measuring oxygen in those ponds. He was on good terms with the farmers. He was an excellent extension person because spending that kind of time, that kind of effort, was just exactly what they needed then."

Kelly took great interest in the farmers' work and was willing to listen to what they were saying and to learn from it. He recognized that the farmers themselves were the ones who went through the trials of managing ponds and raising fish every day. He was keenly aware of the limits of his own knowledge and aware of the collective knowledge and capabilities that existed at his alma mater. Drawing on his university ties to build connections between the farmers and the scientists, he sought to match the farmers' needs with the expertise he knew existed there. The fit was a natural one. During his weeks in Auburn, Kelly nurtured contacts with Auburn fisheries faculty such as Claude Boyd in water quality, Tom Lovell in nutrition, Bill Rogers and John Plumb in pathology, and Wayne Shell.

Knowing they would understand best what was happening at the ponds if they saw it with their own eyes, Kelly invited the researchers to visit farms and observe conditions for themselves. The ensuing discussions were enlightening for all involved. The Auburn faculty learned much about the catfish industry from those exchanges and experiences. Over time, they gained a clearer understanding of the challenges faced by the farmers and determined to try to help. Partly as a result of the visits to farms, AU became more directly involved in the industry's development, and farmers came to understand better what was happening in their ponds and to their fish.

Before there was equipment available with which farmers could diagnose oxygen deficiencies. they had had to rely solely on visual clues, such as noticing whether the fish were swimming slowly around at the water's surface early in the morning or whether they were eating poorly. "It was quite an operation," recalls Buddy Lavender, a long-time farm employee and hatchery manager. Using a little six-volt flashlight, "you drove to a pond, cut your engine off, shone the light to see if you could see any fish around the edges. You got real quiet, and then slapped the top of the truck. If you didn't hear them splash, you went to the next pond. If they splashed, they were up for oxygen."

Many West Alabama farmers learned from Green and Kelly how to use a fifty-dollar chemical test kit. The kit took time but was relatively easy to use. A dissolved oxygen meter that did not require the use of chemicals had been described as early as the 1950s, but the kit was the only practical and affordable method available at the time for most farmers to determine, within minutes, actual oxygen levels in their ponds. Many farmers checked their ponds in the afternoon and again the next morning, but during the hot months of summer and early fall, nightly oxygen checks became routine.

Something had to be done about the problem of oxygen depletion if production was ever going to safely exceed the three-thousand-pound-per-acre limit Swingle had originally advised. Experiments were done at Auburn University on the relationship between feeding rates, ammonia from fish waste, and other water quality variables. Claude Boyd and other researchers were able to determine several feeding rates that would have different effects: (1) a level at which oxygen problems were unlikely to occur, (2) a level at which some depletion would occasionally occur, and (3) a level at which insufficient oxygen would be a constant problem. Yet, relatively little was understood about the specific mechanisms involved in oxygen depletion.

Little was also known about how to get oxygen into the pond. Before the mid-1970s there were few options. From meetings and conversations with fisheries experts, farmers had learned that exposing more surface area of the pond water to contact with the air would add oxygen, so they applied common sense to devise makeshift solutions.

One of the earliest remedies farmers used for oxygen trouble was to run a motor boat around and around in the pond. The boat's propeller created turbulence, and the splashing waves resulted in some aeration of the water. Except for those fish run over by the boat, this method appeared to keep fish alive. Sometimes, one or two weeks or a month later, however, the fish would have more difficulty recovering from a second incidence of oxygen depletion because of the stress they had undergone earlier.

Farmers who did not have a motor to put on their boats tried what was called "rocking the boat." Rocking the boat worked only in small, lightly stocked ponds of no more than four or five acres. The farmer would row his boat into the middle of the pond, stand up in the middle of the boat, and rock it from side to side. The rocking made waves on the surface. Ready to try anything, farmers even backed their Bush Hogs, or tractor mowers, into the pond to stir the water up.

For larger ponds, someone hit on the idea of exposing water to the air using a pump. When low oxygen was suspected, even little five-horsepower pumps were pulled out, cranked up, and kept running all night, pumping a one-and-a-half-inch stream of water out of the pond and letting it flow back in. Electricity was not yet generally available at the fishponds, so the pumps were powered with gasoline. Some farmers would take the pumps into their boat and pump water up into the air through a two- or three-inch pipe, allowing it to fall down onto a board before the droplets splattered back into the pond. They thought that the drops would pick up oxygen on their way through the air. The fish would gather around the newly splashed water, but one or two small gasoline-driven water pumps were not enough to keep all the fish in a pond alive. The farmers felt better, thinking they were at least trying to do something, but it was "like trying to drain a river with a thimble," comments Bill Kyser. Actually, says Claude Boyd, "it prolonged the agony for the fish. It just took them a little longer to die."

Farmers invested money and long, hard hours of labor in any piece of equipment that looked like it might work. A type of compressed air gun that forced air bubbles into water was among methods advertised and reported on in trade journals. Some farmers tried the Crissafulli centrifugal lift pump, that forced large volumes of water up into the air to fall back into the pond.[4] It cost hundreds of dollars, but they found a way to pay for it in the hope that it would save their fish. Some who were able, made their own versions. Experience proved the pump less effective for the larger ponds, however.

In the mid 1970s a company from the East began marketing fountain-type aerators to catfish farmers. Using two- or five-horsepower submersible motors mounted on Styrofoam, they threw the water up like an umbrella. Thinking that the water had to be broken into small particles in order to get oxygen into it, a number of farmers bought them. They installed the aerators in ponds as large as twenty acres, but the machines required too much energy and proved impractical and not very effective. The water they did pick up fell back within a very limited distance of the pump, and the pump tended to pick up the same water repeatedly, having little effect on the rest of the pond.

Farmers pinned hopes anew on each new aeration method they heard about. They reasoned that if aerators put more oxygen into the water, they should be able to grow more fish. As they tried new methods, they also put more fish into the ponds. The result was typically worse than before.

A professor at Oklahoma State University, James Garton, had described in an article a mixing device he had designed called a destratifier, originally used for public water supplies in lakes in that state. The purpose of the destratifier was to mix oxygen-rich surface waters with deeper, oxygen-poor waters. When it was used in the fishponds, the deeper waters, which contained concentrated fish and food wastes and no oxygen, mixed with the upper waters, where the oxygen and the fish were. Dave Kelly recognized potential in the destratifier and spoke with Elvin Wright and other SCS employees. A landowner in Hale County agreed to build one modified for fishponds according to SCS specifications. He put it in one of his ponds.

The destratifier did not provide aeration to the water, but it did on occasion help prevent low oxygen conditions indirectly. At the time, many ponds reached depths of ten, fifteen, or even twenty feet. Kelly had found that plankton populations would grow dense at the deeper levels in those ponds where the fish wastes were settling out, causing an oxygen deficit. The destratifier helped maintain adequate oxygen levels in deep ponds by spreading the plankton populations out, in effect reducing their density and allowing the pond to cope better with their oxygen needs. The construction of shallower ponds helped overcome the problem of stratification, but the destratifier was in widespread use for a while. The machine played an important role in helping catfish production in deep ponds over a major hump until other methods emerged.

Times of low dissolved oxygen meant little sleep and plenty of worry for early catfish farmers. During the summer, farmers and their families often did not see much of each other for long periods of time.

One year, the hot weather lasted for six or seven months. On a steamy August morning, Lenson Montz steered his small boat onto the green, grassy bank, stepped out, and walked toward the house. Turning on the faucet at the side of the house, he washed off his hands. Catching as much water as he could with cupped hands, he gulped it down and splashed water over his face. The warm water ran down into his sweat-soaked clothes.

"Daddy, daddy!" called one of the children from inside the house. "Come and look at what I made!"

"Sorry, sugar, but Daddy can't come look right now. Maybe tomorrow. I gotta go back out to the pond," came Montz's reluctant reply.

"When are you gonna come, Daddy? When are you gonna come look?" asked the little girl.

"I just don't know, sugar, maybe later. The fish are gonna die if I don't keep those pumps goin'," he apologized.

"Now you run along and get some toys to take to Nanny's with you," coaxed Ann Montz, taking control of the situation. "I'm taking the children to Mama's," she told her husband, "and then I'm comin' out to help."

Later that morning, when Ann was back and had changed into some old clothes, she headed out for the ponds. She joined her husband and a second man, a Farm Fresh employee whom Glover had sent over to help when the oxygen was low. All that day, the three worked together, driving the pickup truck from one pond to the next to check the pumps. They made sure the water was still moving, and they greased and oiled the little gasoline motors to keep them running. Ann drove back to the house briefly to make sandwiches for a quick lunch, and then returned. After they ate, she continued to help them as best she could.

During the hot season, it was not unusual for catfish farmers to get only two or three hours of sleep a day for months. What sleep Lenson did get was in the truck, and it was brief at best. That August night, the children stayed with their grandparents. Between rounds checking the pumps, Lenson and his wife tried to nap in the cab of the truck to the sounds of splashing water, katydids and crickets. Because they had to keep the truck windows open, whining mosquitoes bit and taunted them incessantly once the air had cooled.

Ann and Lenson stayed out beside the ponds for three days and two nights. When they made their rounds in the dark, they used the truck lights and a flashlight to find their way. When the moon went behind clouds, the sound of the splashing water helped them spot the pumps and the fish gathered around the boards. They peered into the darkness in front of them as they walked, not wanting to slip in the mud and giving little thought to the snakes that lay among the damp, grassy tufts on the banks. Lenson stepped on several snakes that night on his way between the ponds and the truck, but all he could think about was keeping the pumps running. When a shiny, dark slithering did catch his eye in the circle of light, he would jump back. Whenever he could, he always let snakes go their way, and he never got bitten.

~

This is Lenson Montz's story as he wrote it himself.

> Looking back to the early 1970s, our fish was in dying need of some kind of aeration that would push water up in the air in tiny droplets so it could pick up dissolved oxygen from the air.

At this time, all we had was small pumps that would pump 200 or 300 gallons of water a minute, which was pumped up in the air to fall on a board and splatter, or pumped through small holes drilled in a piece of pipe making it pick up oxygen.

At this time, Mommie and Daddie [J. T. and Murvle Montz] had thirty acres of water. I had sixty-two. We had to have something better.

At this time my brother-in-law Otis Breland was helping us, feeding fish and checking oxygen. It was a twenty-four-hour-a-day job. In the spring of 1975 Otis's cousin, Milton Breland, came for a visit. He gave us the idea that some kind of aeration device could be made by using a truck's rear axles and gear box, mounting it on wheels and taking the tires off and replacing them by welding flat metal to the rim and connecting the gear box to the power unit of a tractor.

A week or so passed. All Otis and I could talk about is building this thing. We needed this thing bad. All fish farmers needed this thing. The whole catfish business—Mississippi, Arkansas, and Alabama—needed this thing. After a sleepless night of thinking about this thing, the next morning I told Otis, "Let's build it." He said okay.

I didn't have much money to spend on it. We didn't even know if it would work. A friend of mine had a small boat trailer. He also had a catfish pond. I asked him if he would sell the trailer. I also told him what we planned to do with it. He said, "You can have the trailer for nothing. I may want to borrow this thing someday."

We went to see Claude Owens. He sold used car and truck parts in Greensboro. He had just what we wanted—a used car back end. Next, we had to have a welder to weld this thing together. Neither of us could weld. We went to Tuscaloosa and bought [a welder], a welding hat, and lots of welding rods. Brought it home and went to work. Our welding was awful. We would weld a piece on; later it would fall off. But we got better at it. We used the boat trailer to mount this thing on, so we could move it from one pond to another. We left the springs on the trailer and mounted the back end to them.

Every spare minute we had, we was working on it, way in[to] the night. I don't know why our wives and children did not run us off. We did not see much of them. But we had to have this thing. After several weeks, we hooked her to the tractor. Off to the pond we went. This was the ugliest thing I had ever seen. But we was proud of it.

We backed it in the water, idled the tractor down and cut the power on. The paddles immediately started turning. This thing was throwing water everywhere. I was sitting on the tractor. I was wet from head to my toes. I jumped off, run around the tractor where Otis was standing. This thing was throwing water everywhere. It was jumping up out of the water, coming back down and throwing water all over the tractor. One wheel would turn a while; it would stop and the other side would [start] turning. Water was going everywhere. There was but one thing wrong—the paddles were turning the wrong way. They was knocking the water out of the pond. Someway we had to make the paddles turn the opposite direction. Nevertheless, this thing would work.

Otis and I both was real proud of it. We both worked at Dairy Fresh Milk Company, making cottage cheese at this time, twelve hours a day.

We went to three mechanics that had shops in town and asked them if we turned the back end upside down, would the paddles turn the opposite direction? They all said, "No." I picked up a pencil, stuck pins all in it. Rotating it in the same direction the paddles was turning and turned the pencil over, and the pens was turning the opposite direction. We took the back end off and turned it upside down and put it back on.

Back to the pond. Backed her in. This time the water was pushed back into the pond, not up on the bank. We still had a problem: one side would turn for a while; it would stop and the other side would start turning. We talked to a neighbor who had a welding shop about our problem. His name [was] (the late) Mike Wurm. He said that was not a big deal. All we had to do was to take a plate off the gear box and weld the spider gears together. He said he has done this lots of times on Model T cars for race drivers. It would give the car positive traction.

We took the plate off, looked in. Otis [asked,] "Which one of these gears is the spider gear?" "I don't know." Otis said, "It must be those two." We started welding [and] the oil caught fire. We got it put out. Put the plate back on. Back to the pond. This time both paddles would spin at the same time. This was good.

But we still had a problem. If we gave the tractor more gas, it was just jumping almost out of the water, up and down.

Back to the cutting torch and welder. We cut out the springs and welded it back. Back to the pond. This time it just sit there and knocked water just like it was supposed to. We was so proud of it, we went to town and bought a gallon of metallic blue paint. Parked this thing in Momma's yard under the pecan tree. Gave Momma a paint brush. When we got off work the next day, she had it painted. It was the prettiest, ugliest thing I had ever seen.

People passing on the road would stop and look at it and wonder what it was. Some would even stop and ask, "What is that thing?" We had this thing for about a year.

One Saturday morning, my father-in-law, Aubrey Morrison, called and asked if a friend of his, Judge Hal Knight, could borrow this thing. His fish was dying from lack of oxygen. We said, "Yes," and we would carry it to him. On our way, we had to go right through town. Lots of people was in town that day. As we came down Main Street, people would stop and look at this thing. Some store owners came out, as we stopped at a red light, and asked, "What is that?" We was proud of this pretty blue thing. We carried it to Mr. Knight's pond, hooked it to his tractor, backed it in the water. Soon, his fish had stopped dying from the lack of oxygen. We left the thing running.

A few days later, Mr. Knight had a fish doctor (O. L. Green) looking at his fish. Mr. Green saw this thing running and was impressed with what it was doing. He passed on what he saw to different fish farmers.

The phone started ringing. Farmer[s] wanted to look at this "thing." They wanted to buy it—it was not for sale. They took pictures. They measured it this way and that way. They came from Mississippi, they came from Arkansas and all over Alabama.

We finally had to back it in a pond and leave it running for them to look at.

Auburn University made one and said it was the best thing they had seen to put oxygen back in the water and to put a current of water all over the pond. This thing later was called the paddle wheel.

The paddle wheel has been a real asset to fish farming. Today, every fishpond has one to three electric paddle wheels in it. Today, electricity is used for power. Tractors are used for emergency.

Thank you, Otis. We built it, didn't we?[5]

Elvin Wright answered his phone at the SCS office. Lenson Montz was on the line.

"Hey, Elvin. Lenson here. Can y'all come by my place today?" Montz asked. "I've got something here I want you to look at."

"Hey there, Lens. Well, I can come sometime later this afternoon, say, around 4:30. I'll be by about then," came Wright's reply.

Later that day, Wright turned his truck off the highway onto the dirt road that led to Montz's house. He drove around the large fishpond and parked in front of the house under the large, old oak tree. Montz came out the door and strolled over to Wright. He was not one to get in too big a hurry about anything. The men shook hands in greeting.

"Come on, let's ride down and look at one of the ponds," invited Montz.

"Okay."

The two men drove down to a fishpond and stopped at the bank next to an old Allis-Chalmers tractor with something behind it.

"What in the world you got there?" asked Wright, looking at the odd thing behind the tractor.

"I don't know," replied Montz, not too sure himself. They arrived at the pond bank and climbed down.

"You stay right over here. I'm going to back that ol' tractor into that pond and show you just what I've got, and *you* tell *me*." Montz backed the tractor partway into the pond and threw the contraption in gear. It started kicking up the water. It looked to Wright like it was doing a good job of it, at that.

After they stood there watching it for a while, Wright asked, "Have you checked your oxygen behind it?"

"I tried to," replied Montz. He had thought of that and had taken his test kit to see what effect, if any, the machine might have been having. "It kicks it up two parts, probably three maybe, around it." Montz was referring to parts per million, or ppm, the measure of concentration of the oxygen in the water. His test kit had indicated a concentration of between two and three parts of oxygen for every million parts of water in the aerated part of the pond immediately behind the machine.

"Man, that's good!" exclaimed Wright, the wheels in his mind starting to spin. A big smile spread across his face. Could this really be? "That's great!" He was still not sure whether to believe what he was seeing. Thinking while he was talking, he said, "Then it didn't get it down, so if you stirred any muddy water or any thing. . . . Then it's just taking the surface water and. . . . That looks good, Lens! Did that thing work the first time you tried it?"

"No, sir!" replied the farmer, shaking his head and remembering his trials of the past few weeks.

"What happened?"

"Damn thing like to drown me!" Montz declared. "It throwed the water back over on *me* instead of throwing it back toward the *pond*. I didn't know that thing would throw the water back on me! Well, I just turned the thing over—bottom side up. We did it at the shop. And now it throws the water back over the pond. What do you think of that?"

"Lens, I believe you've got something here," came Wright's understated reply.

Wright had seen the first paddle wheel aerator for a large pond that was ever built, and he knew it was going to work. It was something that every fish farmer could build and use. He passed on the great news.

"Go look and see what ol' Lenson's got," he told all the farmers he knew. "You can get a little ol' truck back end, and you can build this thing in the shop. You can build it for *you*. And you can build it *yourself*. You've got tractors you could hook it on and take it out there to the pond."

The news spread like wildfire, and the paddle wheel caught on quickly. Montz and Breland had demonstrated a way to put the surface water of a fairly large lake effectively into the air and to create a current that brought deeper, oxygen-depleted water up to be aerated. Even though the fish were distressed, they managed to come in and stay in the large wake of the paddle wheel until enough oxygen had been eventually added to the pond for them to be sustained in other areas without stress. The SCS office phone rang even at night with people inquiring where they could go to see the contraption Montz had built (see fig. 11).

The catfish industry started out in Alabama as a do-it-yourself and figure-it-out-yourself kind of enterprise.

"If you needed something you had to build it or have it built," says Bill Kyser, because no suitable equipment was available. The invention and development of the paddle wheel continued that pattern. Many of the farmers had taken metal shop in high school, and knowing how to cut and weld metal, they built them for themselves. Some with larger farms who employed mechanics of their own had employees build them. Others had local machinists build custom-made paddle wheels for them.

Now that a device had been found that worked and was relatively simple and inexpensive to make, people got ideas for improvements. Farmers and mechanics made small adjustments on their own versions. Some tried to make more powerful paddle wheels and ones that would float suspended from Styrofoam blocks out in the middle of the ponds.

The machinist Clark Livingston had been contacted by Lenson Montz to help get some parts for the original paddle wheel he and his brother-in-law had been working on. Livingston had gone by the Montzes' farm to see the finished product at work. Shortly afterward, he received an order for

Fig. 11. Lenson Montz standing next to reconstructed paddle wheel, 1998. Montz and Otis Breland, his brother-in-law, built this reconstructed paddle wheel using some of the original parts, the rear axle of an old truck, and an old boat trailer. (Photo by author.)

one from another local farmer and, over the next six or seven years, built more than fifty paddle wheels for various farmers in the area. Around 1980 he built and delivered twenty-four to Paul Smith at ConAgra's large farm in Mississippi. Livingston's focus was on keeping his small machine shop business viable, however, and he never built up enough inventory to start a large-scale business. That niche was to be filled by others.

With time and experience, farmers and others learned how to best use and modify the paddle wheel. Occasionally, learning came after disaster, when a farmer would find that using the machines only intermittently did not prevent fish kills. Farmers did not know the paddle wheel had to run continually to keep fish in heavily stocked ponds alive through very hot days and nights. In addition, they did not realize that the device did not need the full power of large tractors to operate.

Before long, the research capabilities of Auburn University were drawn into the development of this new method of aeration. Claude Boyd had been researching aquatic plants and water chemistry since he joined the faculty in 1968. By the early 1970s he was receiving calls from Mississippi catfish farmers asking for help with oxygen problems. His education and training had prepared him to work not on biology but rather on chemical and physical problems, and he was being called to apply that training to biological problems with pond-raised catfish. No one else at the time appeared to know or to want to work with such questions of water quality, so he picked up the baton.

Researchers had contributed to the mainstream of aquaculture development, but they had not investigated the topic of water quality. Though destined to become a core issue in commercial catfish production, aeration research at the time was difficult to fund. Conventional funding sources such as government agencies did not approve the projects Boyd proposed. Most research was funded, and still is, according to Boyd, on the basis of whether it appeared to be something new, whether it was basic research rather than applied. Aeration as such was not new. The problem was well known from issues of municipal water supplies. Yet it was a real bottleneck in the catfish industry.

Auburn University had by that time developed an international training program in aquaculture. Not only did foreign students come to receive fisheries training, staff received funds to work on development projects in other nations. These activities were paid for by funds from the U.S. Agency for International Development (USAID), which provided core support programs and strengthening grants for Auburn's fisheries research. The head of Auburn's fisheries program, Wayne Shell, finally found a way to pay for Boyd's research through USAID funds. The producers' organization, the Catfish Farmers of America, also contributed.

Theory and academic research were less Boyd's concern than working on the actual problems farmers were having. Once the decision was made to work on those problems, Boyd went out to the farms to find out what the farmers wanted or needed. After he learned about the paddle wheels, he began the first systematic study of the paddle wheel method of aeration. He and fellow researchers designed a prototype on which various parts, such as the paddles, and the speeds at which the wheels turned could be changed and the different results determined (see fig. 12).

The research itself did not pose particular problems. There were just thousands of tests to go through and many different combinations to try. It was clear that a few factors affected how the paddle wheels operated, such as how deep the device sat in the water, how fast the wheel turned, and the

Fig. 12. Experimental electric paddle wheel, 1986.
Auburn University researchers developed an experimental electric paddle wheel for testing and improving the design of electric aerators. The position of the paddles is adjustable to determine the most efficient arrangement. (Used by permission of Claude E. Boyd.)

shape and position of the paddles. From the many combinations of factors that were tried, the purpose was to find the one that gave the best result in terms of oxygen produced and efficiency of operation.

Boyd was no mechanic, but he observed that tractors were not made to run at the angle at which they sat when backed down the bank of a fishpond. He judged that because of gravity, the oil would inevitably run back into the motor. He observed a machine that had cost a farmer thousands of dollars being used for powering only one aerator. When he started seeing tractors accidentally backed *into* the water, he noted the impracticality of moving them around between ponds in the middle of the night. Clearly, there were many problems with using tractors, and it became obvious to him that some type of floating electric aerator would be much better.

Thed Spree could not have agreed more. Spree had thirteen tractors and more than thirteen ponds. Every morning, he had to fill the tractors with fuel, grease the paddle wheels and the tractors, and change the oil. He had to pull the paddle wheels out of the ponds and let the water run out of the back ends. He had to go look for this part or that, a ring gear or a pinion, and put the motor back together before the next night so that he would not lose his fish. Entire days were spent just greasing paddle wheels, checking tractors, and moving them from one pond to another.

~

The final product coming out of the work done at Auburn University was a result of many people's combined experience and knowledge—researchers, farmers, mechanics, and others. The mechanism they developed turned out to be the key to overcoming one of the catfish industry's biggest stumbling blocks. Soon the entire industry was depending on what had started as an "ugly" contraption put together by a catfish farmer and his brother-in-law, two men who had barely any idea at the time what they were doing. They had known only that they were desperate to keep their fish alive. The paddle wheel has since become arguably the most widely used method of aeration in the commercial culture of numerous fish and even some shellfish species around the world. It is produced and sold by manufacturers internationally.

Neither Lenson Montz nor Otis Breland ever applied for a patent on the "thing." But, as Ann Montz recalls, there is great satisfaction in knowing they contributed to the success of the catfish industry.

> The children and I could look at Daddy's face and know immediately if "this thing" was working properly. Time and time we saw in his eyes that it wasn't. We saw how tired he was, how worried. We came to realize how really important this was, and we are so proud that he kept trying—never gave up. And we were rewarded when it finally all came together—with a smile a mile wide and relief that "this thing" was the answer to the most important problem the fish industry faced. And when we ride and see wheels turning in every pond, we know. We know![6]

Notes

1. Jerry Crews, interview by author, tape recording, 19 February 1998, Special Collections and Archives, Auburn University, Auburn, Ala.

2. Richard T. (Tom) Lovell, interview by author, tape recording, 15 April 1998, Special Collections and Archives, Auburn University, Auburn, Ala.

3. Generally, the feed conversion ratio is a way of expressing the relationship between the weight of feed offered and the weight gained by the fish as a result of

that feed. Thus, a ratio of 2.0 (2:1) indicates that two pounds of a feed bring about one pound of growth.

4. Stepanoff gives the following definition of centrifugal pumps: "Centrifugal pumps comprise a very wide class of pumps in which pumping of liquids or generation of pressure is effected by a rotary motion of one or several impellers. . . . [The impeller] forces the liquid into a rotary motion by impelling action, and the pump casing . . . directs the liquid to the impeller and leads it away under a higher pressure" (*Centrifugal and Axial Flow Pumps* [New York: John Wiley and Sons, 1957]), 19.

5. Lenson Montz, "Looking Back—1996," 1996, Special Collections and Archives, Auburn University, Auburn, Ala.

6. Ann Montz, "A Note from One of the Wives Who Lived through This Story," 1996, Special Collections and Archives, Auburn University, Auburn, Ala.

12
Fish Health

> The commercial production of catfish has "trouble" written all over it when you consider that it is high-intensity fish culture. Growing fish under such crowded conditions greatly increases the chance of diseases. If the water quality deteriorates, then the fish become stressed and more subject to diseases. There's a logical limit—a law-of-nature limit—on how many pounds of fish a given body of water will support. These farmers were mostly ignorant of these limits. They were going to produce these fish regardless of what anybody said.
>
> —Dave Kelly

Inadequate oxygen was not the only difficulty catfish growers faced in keeping their fish alive. Disease and parasites posed threats in all phases of fish farming. Clearly, if commercial fisheries were to succeed, ways would have to be found to keep the fish in the hatcheries and the ponds alive.

In 1954 Ray Allison at Auburn was one of only three fisheries parasitologists focusing on fish health problems in the United States and the only one working on parasite problems of warmwater species native to the Southeast. Researchers had observed warmwater fish epidemics years earlier among populations in natural lakes, waters, and ponds. Examining dead fish under a microscope, Allison saw that their bodies were covered with protozoan parasites. Identifying parasites of wild warmwater fishes thus became an initial focal point in the fight against disease. The number of parasite species seemed endless. For many years, researchers found at least one previously unknown parasite on virtually every fish they netted in the streams.

During the early days of commercial catfish farming, farmers relied frequently on wild fish they caught in the rivers for broodstock. Inevitably, the fish brought parasites with them, and some would transfer to other fish in the ponds. The typical life cycle of the parasites involved eventually dropping off the fish, settling into the soil on the pond bottom, and later, reattaching to different fish. This cycle was hard to break.

The protozoan parasites *Trichodina* and *Ichthyophthirius* were the most common and caused the most problems for commercial catfish farmers.

Ichthyophthirius, or Ich, was likely the most devastating parasite at the time. It was, and still is, very difficult to treat successfully in ponds. The fish could also harbor roundworms, flatworms, flukes, and other internal parasites.

Parasites were only one source of fish health problems early growers encountered. Infections by the *Aeromonas* bacterium and the myxobacteria that caused columnaris infections were just two of the many other organisms that killed pond catfish. Fungal infections commonly developed on eggs in hatcheries and could be responsible for problems in grown fish as well. Occasionally, in STRAL's early years, egg masses in the aquaria would "fungus up" even before breeding pairs had finished spawning.

Many parasites causing problems lived on the outside of the fish's bodies. That made them susceptible to almost any treatment that could be applied to the water in which the fish lived. During the last half of the nineteenth century, certain procedures had been developed for treating diseases in aquarium fish. Those procedures involved chemical treatments into which the fish were dipped. Two of the chemicals used were methylene blue and malachite green. Another early dipping treatment for parasites used common table salt, sodium chloride.

Commercial trout culture, which preceded catfish culture in the United States by roughly a hundred years, was typically done in small concrete tanks or raceways.[1] To treat infections, the fish were lifted out, dipped into a container of prepared solution, and returned to the tanks. These chemicals had not yet been tested in pond culture, and commercial catfish farmers had few methods of treatment to which they could turn. How was one to treat fish that swam freely in ponds of one-half to ten or more acres in size and containing, at the *very* least, thousands of gallons of water?

Part of Ray Allison's job was to find a way to do that. At first he used chemicals, any at all, looking for levels of concentration at which solutions would kill a parasite on a fish without killing the fish itself. Because of concerns about spreading the parasites or disease organisms, Allison could not use the ponds at the university's research facilities. When hatcheries or growout farms in nearby areas faced parasite or disease outbreaks and the growers asked for help and gave their permission, he was able to test his ideas there. These hatcheries and farms were the primary laboratories in which methods and guidelines were developed for the use of malachite green, potassium permanganate, formaldehyde, and salt on pond fish.

Eggs in STRAL's very first hatching troughs had been destroyed by fungus. The partners had sought advice from experts at the federal hatchery in Marion, where biologists worked with a variety of warmwater fish. At that time, the dye malachite green was known to act as a fungicide, and it proved effective at the hatchery. Thus, even in the early years, hatchery operators began routinely dipping egg masses, barehanded, into plastic dishpans full of the dye as soon as they took the spawns away from the males, as a preventive measure. Greensboro residents knew when it was spawning time by the STRAL partners' bright green hands and arms, and the stains on their clothes. Malachite green proved effective also on a number of parasites, including Ich.

~

By the 1950s the fisheries program at Auburn had begun to shift its research emphasis from recreational fishing toward commercial fish production. Homer Swingle was convinced of the huge potential that fisheries held for food production in countries with undernourished populations. He traveled abroad extensively and brought back ideas for ways in which he and his faculty might apply their knowledge to promote fish culture in developing nations. In the late 1960s and into the 1970s, with USAID funds that supported the faculty's research in fish production and other aspects of fisheries development, Swingle began hiring people with Peace Corps experience. Eventually, approximately half or more of the graduate students and faculty were returned Peace Corps volunteers, providing the department with a unique storehouse of skills and knowledge.

Their international experience made the faculty's orientation toward their jobs very practical. Overseas work, often characterized by few resources and challenging conditions, heightened their resourcefulness and enriched their ideas and approaches to solving problems. Their expertise worked ultimately toward the benefit of the catfish industry, because the department's research focused more and more on catfish as the "test" fish. While faculty members were paid to be on call for assignments abroad, when they weren't traveling they worked with visiting professors and students on catfish culture, physiology, nutrition, disease, and genetics; pond water quality; and related problems. This knowledge and experience provided the foundation on which the industry still grows today.

~

Beginning in 1950 state biologists could learn more about fishing lake management by participating in trips led by Homer Swingle to various state-managed lakes in Alabama. The tours were sponsored by the Fisheries Section of the Fish and Game Division of the Alabama Department of Conservation and Natural Resources. During one week in June, they, along with fisheries faculty and students, would travel to two or three such lakes every day

Fig. 13. Homer S. Swingle lecturing during state lakes tour, 1964.
Homer S. Swingle talks to a group of fisheries biologists, possibly at Tuskegee City Lake. Rudy Schmittou is on the left. (Used by permission of Auburn University Libraries, Ralph Brown Draughon Library, Special Collections and Archives.)

to check the population balance and the conditions of the aquatic environments in each. They would then discuss and learn from Swingle's lectures on the results of their investigations (see fig. 13).

Later in the summer, in August or September, employees from southern states' fish and game departments joined in annual fish population surveys, also led by Swingle, at the state's rivers and reservoirs. Often, forty or fifty people participated. The group would camp out for about four days next to a river or a lake and identify, count, and measure fish they collected (see fig. 14). The data were used to determine the species' distribution, the relationships between the different species, and the condition of the fish.

The participants had serious work to do during the lake tours and camp-

Fig. 14. Fish population count, possibly early 1950s.
Fisheries biologists and students conduct a population count on one of the many fish population study trips conducted around the state. The studies were done under the auspices of the Alabama Water Improvement Commission before Homer S. Swingle (*seated in shade, background far left*) became interested in aquaculture. (Photo courtesy of the Department of Fisheries and Allied Aquacultures, College of Agriculture, Auburn University.)

ing trips. Mortalities in sport fish hatcheries and fisheries were common throughout the Southeast. The groups' task was find ways to combat the problems by building on the data they collected and on members' knowledge and understanding of fisheries and aquatic biology. Out of the discussions in cars, motels, and around campfires came Swingle's idea of a cooperative diagnostic project among the states. If each state contributed a small part of the cost, he reasoned, tremendous benefit could result for all of them from the services AU could provide.

With briefcases, slides, projector, and pointer, Swingle, Allison, Wilmer (Bill) Rogers, and others visited state fisheries officials and presented their proposal. The response was enthusiastic, and acceptance was immediate. Initially, twelve states took part.

The Southeastern Cooperative Fish Parasite and Disease Project (the

Project) was just one of Swingle's innovations. The Project began in 1964, the first of many years of service to southeastern fisheries and the catfish industry. With Rogers, who was primarily a parasitologist, as director, the knowledge and efforts of resident and visiting faculty at Auburn were brought into play to solve disease problems in the region's warmwater fisheries. The Project drew on the assistance of the many students coming to Auburn for graduate degrees. Sometimes as many as fifteen at one time pursued some aspect of the Project's work for their thesis or doctoral research.

Every year, researchers presented seminars in one of the participating states. They trained specialists in fundamentals and new findings regarding parasitism, nutrition, water quality, and related areas. Those specialists were then able to do routine diagnostics in their own states. When they needed help with specific problems, they sent samples of affected fish to Project staff, who performed tests and got the results back to the inquiring state as quickly as possible. At the beginning, many of the diseased fish were from natural waters such as reservoirs, and nothing could be done for them. At least, however, a cause could often be determined, whether it was disease or pollutants, and that information added to the total store of knowledge.

Within just a few years after the Project was established the emphasis began shifting increasingly to diagnoses involving channel catfish. With growth in the industry, the caseload grew until catfish composed 70 or 80 percent of the total three hundred to four hundred cases diagnosed each year in the Project's laboratory. In the process, the faculty and students identified and described approximately 150 new parasites and diseases. They also developed diagnostic procedures using fluorescent antibody and immunoassay techniques, tests that confirmed findings more quickly than previously possible.[2] These procedures made it easier to identify organisms and possible treatments early enough to prevent great losses of farmers' fish.

If catfish farmers had some disease or mortality among their fish, they knew they could call Project staff and not be turned away. If there was a solution, the researchers would know about it. Most farmers generally tried to save the long-distance call to Auburn as a last resort, however, only after their own solutions and waiting and watching had failed. To Project staff, that was like waiting until the ninth inning with the bases loaded. Chances of detecting the culprit in time to effectively prevent a fish kill were greatly diminished by then.

Professors and students who worked in the lab put in long hours and weekends. During the week, someone was always present during work hours.

Evenings and weekends, they took turns being on call and were available, if needed, to head over to the Greyhound bus station, the post office, or the university to receive sick fish. They tried to get farmers to bring live fish to them, if possible, but Greyhound service was otherwise the cheapest and fastest way the fish could be sent.

Bill Rogers arrived home one Friday after a long day at the laboratory. He looked forward to a satisfying meal and an evening with his family. No sooner had he walked in the door, though, than the telephone rang.

"Bill Rogers here," said Rogers, hoping against hope that it was not a fish disease emergency.

"Hello, Dr. Rogers? Thomas here, from the Greyhound station. We have another one of them smelly, wet packages for you again, came in on the last bus. From Georgia this time. Would you mind coming by and picking it up?"

"Thanks very much, Thomas. I'm sorry. I'll be right over," Rogers replied. It could have been worse. Some of the calls came in the middle of the night. It could have been a worried farmer wanting a diagnosis and advice over the phone. At least the package had not been misaddressed like the one delivered last week to the office of the dean of agriculture instead of the Fisheries Department. Project staff had had to go and pick up the dripping, reeking package from the dean's office. The diseased fish had been shipped live, sealed in a can. The contents had been totally liquefied by the time the box had arrived several days later and were useless for diagnosis.

Project staff had a difficult time conveying to farmers, and to fishery biologists in general, how to get fish to them in a condition suitable for examination and necropsy. They simply could not use fish that had been dead very long, because the fish began to decompose almost immediately. And so when people would call, the researchers would request the fish be preserved in formalin or frozen, on dry ice. Once or twice, they received formalin-preserved fish that had been shipped in glass through the mail, and the glass had broken. Post office employees knew immediately what was in the package and would call the laboratory and politely, but urgently, request, "Please come get this box of fish!" When Rogers or John Plumb opened such a package, clouds of fumes would burn their eyes. These occasions prompted the researchers to put together specific instructions for farmers on how to send fish and how *not* to send them. After those early awkward and unpleasant experiences, farmers eventually learned how to get fish to the lab in usable condition.

Many Alabama farmers brought their fish in personally. Sometimes, but not always, they called ahead of time. Sometimes they just walked in the door of the laboratory carrying a metal bucket or other container with a sick fish at their side.

John Plumb joined the faculty in 1969 as a virologist, bacteriologist, and general microbiologist. His job at the time was that of disease diagnostician. One cool fall afternoon, while working at his office desk, he looked up to see a farmer with a bucket standing in the doorway. Plumb got up from his chair and walked around to shake the man's hand. After introductions, he led the man to the parasitology laboratory, which at the beginning had been nothing more than a 5-x-8-foot "clean" room built within a larger room. In 1974 a building specially designed and built for the needs of the fisheries program provided the Project with a separate lab of its own. Along the way, Plumb asked the farmer a series of questions like the following: "How long ago did you first notice something was wrong?" "Are you seeing more and more sick ones every day?" "How big is the pond?" "How many fish do you figure you've got in there?" "How many do you figure you've lost so far?" "How are they acting?" "What have you done so far?" "Have you treated them with anything?" The farmer was very cooperative and tried to give as much information as he could.

More questions came: "What's the water like in the pond?" "Do you know how much oxygen you've got? how much CO_2?" "What's the water temperature?" Plumb was checking to see whether water quality might be at fault rather than a parasite or disease organism. If the fish had died all of a sudden the night before and the whole pond lost that morning, he could be fairly certain that was what it was. However, if a couple of fish had died the previous week, and the number of deaths had been increasing since then, some sort of disease problem was likely.

Once in the lab, Plumb took a white lab coat off a hanger and put it on. When he flipped a switch, the laminar airflow hood over the blacktopped lab table began to hum. The hood was filtering air from the room before it passed the "sterile" air down over the workspace below it where Plumb would be examining the fish. He wanted to be certain no dust, bacteria, fungus spores, or other air-borne contaminants would come in contact with the fish's body. He washed his hands over a small sink with an alcohol solution. Preventing contamination of the test equipment and the fish itself by anything the staff brought in was crucial, of far greater concern than the possibility of their contracting anything from the diseased fish. He turned on a lamp that shone onto a sterile tray on the table. With his clean, bare hands, he lifted the fish carefully out of the farmer's bucket and laid it on the tray under the light. The farmer watched with interest.

Having concluded that water quality was not an issue, Plumb began examining the exterior and the gills of the fish. The skin was not quite grayish in color, and it looked almost as though someone had sprinkled salt over it. Little white balls dotted the body and fins, just under the outer layer of the fish's skin. It did not take long to conclude that this fish was in the early stages of an Ich infection. The Ich parasite was particularly difficult to remove from ponds.

Instead of stopping at an initial diagnosis, Project staff did a thorough examination. With catfish, they always did parasite, bacteriological, and virological assays. Even when they saw a parasite problem, they did the bacteriology and virology, because there was always the possibility that more than one disease was present.

The fish had to be dead, however, before Plumb could complete the examination. In those days, "pithing" was a diagnostically acceptable method of killing the fish. The procedure severed the spinal cord and killed a fish almost instantly. Plumb and the other researchers had learned early on to ask permission from the owners to pith the fish.

"I'm going to have to kill your fish. Do you mind if I do that?" he asked the farmer.

Many of the disease problems occurred in ponds with thousands of individuals in them. Farmers were prepared to dispatch a few to find out what was wrong with the multitude. The farmer gave his okay.

To complete the parasitology exam, Plumb made slides from the lesions and looked at them under the microscope. What he saw confirmed his initial diagnosis. He invited the farmer to take a look. While the man peered awkwardly through the lens and tried to focus on one of the parasites, Plumb described for him the nature of the Ich organism and infection. With a definite parasite identification, he was able to tell the farmer immediately what he might be able to do to save as many fish as possible.

A bacteriological assay, however, would take at least a day, and maybe two or three.[3] Plumb would ask a student assistant to isolate the bacteria and let them grow in glass Petri dishes so they could be identified. The dishes would be kept in one of the small, refrigerator-like incubators in the lab where cultures from warmwater fish were kept. If bacteria were found, the researchers would want to determine whether or not the organisms were antibiotic-sensitive. That would take two or three additional days. If they were, the farmer would have a reasonable chance of correcting the problem with an antibiotic.

The virological work would also take several days before it offered any results. A virus sample had to be taken from cells in the tissues of the fish and grown in tissue cultures. Everything had to be sterilized, so none of the cultures would be contaminated with bacteria.

Sometimes Project staff would take pictures of the fish as they were examining it. They wanted to record the clinical signs and appearance of the diseased areas. They would wrap the fish and place it in a big freezer downstairs in case they might need to look at it later. With no unusual signs or symptoms present this time, Plumb decided to forego these measures.

Plumb wrote down the farmer's telephone number. He knew from previous experience how crucial it was to get results to farmers as soon as possible so the farmer could try to prevent catastrophe in his ponds. Plumb assured the farmer he would call him as soon as he had the results, and the farmer left.

In the early years of the industry, fisheries experts were learning right along with the farmers. Being a generally realistic group, most of the farmers understood this and were receptive to the advice they received. They had a good idea of how many fish they had already lost, and they realized they were going to lose more. They were understandably not happy about it, but they did want to find out what was going on and whether anything could be done. Like most other farmers, once they learned something about the biology and ecology of their crop—and, in this case, the complexity of the aquatic environment—they understood that many things could interact to cause mortalities. When it was disease, the culprits were often of more than one kind, typically both parasites and bacteria. It could also be water quality, mishandling, or some other poor management practice.

When both parasite and bacterial infections were diagnosed, the first decision to be made was which of the two to treat first. The question of cost was a major one, and the expense of the treatment had to be weighed against the value of the fish the farmer was likely to lose. Sometimes the farmer and Project staff could determine whether treatment was worth the cost. Sometimes they could not. Often the matter came down to a gamble of sorts: whether to put many thousands of dollars into treating a large volume of water containing large numbers of fish when perhaps only a limited number of the fish would succumb to the infection. The decision was not always an easy one.

There were times when Project staff started out thinking they needed to diagnose diseases and ended up instead trying to figure out where farmers had gone wrong with treatments they had applied. Having invested hundreds or thousands of dollars worth of feed and months of labor into a pond, the prospect of losing all of it to Ich or some other parasite or disease inevitably struck farmers with dread. In their fear, some applied too much of a certain chemical and learned the hard way that that approach had only caused another problem. Sometimes, farmers who wanted to kill off weeds applied too much herbicide, such as copper sulfate, and found that the herbicide treatment had been toxic to the fish or had totally cleared their pond of the plankton needed to provide the fish with oxygen.

Bill Rogers was met in the lab one day by a farmer who had brought some catfish in for examination. Rogers found the gills ragged and bloody.

"They've got gill melt," stated the farmer, trying to be helpful.

"How did you determine the gill melt?" Rogers asked. He had never seen or heard of this condition before.

"Well, one of the fellas who's got ponds near mine saw it and told me it was gill melt," came the reply.

"Have you done anything for it so far?" asked Rogers, running through the usual questions.

"I've been treating it with potassium permanganate."

"How much potassium permanganate have you been using?" asked Rogers, thinking.

"I reckon about two or three parts per million or something like that, every day or every few days or so."

That confirmed Rogers's hunch. "Well, I think that might be your problem right there," he said. "You may be burning those gills with potassium permanganate."

The farmer left with instructions to stop the treatments for the time being, and Rogers sat down to plan a small experiment. He treated some catfish in a tank with two parts per million of potassium permanganate. The next day he did it again with the same fish, and again the following day, and so on. After four or five days, he stopped the treatments. The fish had started dying, and their gills were ragged, bloody, and perforated. In his effort to rid them of disease, the farmer had been inadvertently killing his fish.

It would take a while for many catfish farmers to learn enough about diseases and treatments to know not to treat their fish to death.

The activities of the Project were in full swing by the late 1960s. A variety of chemicals and antibiotics were being commonly used to prevent or treat a number of infections in the eggs, fingerlings, and grown fish. While the industry was still very young, little or no testing of chemicals had been done to determine the safety of their use either on the fish or on the people treating them, much less on the people who ate them. No policy or regulations existed that limited the use of any chemical in aquaculture. Few people gave much thought to aquaculture's impacts on the environment or on consumers. Because of the lack of regulation, the field was wide open.

Oxytetracycline, known as Terramycin, had been discovered in 1949 and had been approved for traditional livestock such as cattle and sheep but not officially approved yet for food fish. The antibacterial drug nitrofurazone was another effective tool in fighting diseases. Copper sulfate helped control disease by eliminating algae blooms. Copper sulfate and other herbicides controlled parasites effectively, and by the time the catfish industry got a firm foothold, those chemicals, malachite green, and others were used fairly widely. In 1970 farmers received a powerful boost in the fight against bacterial infections when Terramycin and certain sulfa drugs received final official approval from the U.S. Food and Drug Administration (FDA) for use in food fish. After that, Terramycin began being added to feeds as a preventive or control measure. Additionally, the establishment of good sanitation in the hatcheries significantly decreased the incidence of parasitism on fingerlings and their transfer with the fingerlings into growout ponds.

Gradually, as the public became more aware of possible detrimental effects of these treatments, federal agencies began conducting and sponsoring research. Results suggested that these chemicals were potentially harmful either to the environment, to the producers coming into contact with them during treatment, or to the people who ate the fish. Substances such as malachite green and nitrofurazone, for instance, were found to be carcinogenic. As the 1970s progressed, the FDA began clamping down on their use. The numbers of therapeutic agents began to drop as, one by one, they were declared unapproved for use in aquaculture. By the late 1990s only a few were legally available for treating diseases in food fishes—primarily only sodium chloride, common table salt, for use on the fish and iodine, or betadine, as a dip for the eggs.

Around 1966 Project researchers began hearing reports during the summertime of fingerling catfish dying by the thousands. Nikola Fijan, a visiting scientist from Yugoslavia, began to investigate. He and others began growing cultures from samples of diseased fish. They were eventually able to isolate and identify the culprit. A virus was causing the catfish kills.

After Fijan left Auburn, John Plumb joined the faculty and began working on the channel catfish virus problem. He spent hours in and out of the lab and out at ponds, learning how to identify the disease in live fish. Everyone was very concerned that the virus would prove devastating to the catfish industry, because when it hit a pond of fingerlings, as many as 80 or 90 percent of them died. In time, the researchers learned that the disease had been around for some time but had not previously taken such a disastrous toll. The problem seemed particularly acute when farmers were crowding their fish.

Laboratory experiments helped the scientists to better understand the disease. A herpesvirus residing somewhere in the fish's tissues emerged to cause havoc. Although a cure has still not been found, the disease never developed into the catastrophic problem originally feared. It remains rather like "being struck by lightning," according to Plumb. "If a farmer has channel catfish virus in juvenile fish, he'll very likely lose 50 to 70 to 90 percent of them, so he's concerned about it. But if he doesn't have the problem, he isn't really too concerned."

~

When Tom Lovell first came to be AU's fish nutritionist, he knew virtually nothing about how to keep any fish alive, much less channel catfish. His experience had been with ruminants such as cattle and sheep, and although he had worked a little with seafoods, he was starting basically at the ground floor with fish health. Within a year or two, farmers began bringing fish into the diagnostic lab with a peculiar growth anomaly. The fish looked almost as if they had had invisible rubber bands around the base of their tails causing their backbones to be bent. Researchers could find no pathogens responsible for the disease. They named it "broken back syndrome." The cause was a mystery.

Swingle had a hunch that the disease might be caused by a nutritional problem and asked Lovell to see what he could find out. He gave Lovell a few resources to set up a lab, some tanks for growing fish, and a small budget to formulate some experimental diets. The problem was now in Lovell's hands.

Lovell uncovered the first clue when visiting and talking with farmers who had seen the syndrome in their fish. He noticed that the fish in question were in most cases in cages. The fact that cages afforded the fish very little access to natural foods suggested to Lovell that something might be missing in the fish's diet. He took a standard nutritional research approach—the purified diet: feeding fish a diet with all the essential nutrients except one. Because the backbone was deformed, he tested for minerals first. The fish all developed normally. Next he fed diets lacking in each of several vitamins. The fish showed some abnormalities but did not develop broken back syndrome.

Checking the scientific literature, he found that a few types of animals, primarily humans and guinea pigs, have poor bone development if they lack enough vitamin C. Lovell prepared diets with all the nutrients except vitamin C, and shazam! The fish developed broken back syndrome. The entire project took close to three years, but the researchers finally had empirical proof. They then fed commercial-type diets with and without vitamin C to catfish growing in the university's research ponds in cages, where they could get no natural food. They got the same results. Knowing that vitamin C was easily affected by oxygen, Lovell and nutrition researchers measured the effect of the extrusion process on the feed during its manufacture in the mills. They found that in most extrusion mills, 40 to 60 percent of the vitamin was lost. So they recommended that manufacturers add double the amount needed by the fish.

This was important information for the catfish industry, but more had to happen till the industry actually benefited from it. Findings were written up and reported at meetings and at workshops of industry groups like the Catfish Farmers of America. The news spread among extensionists and farmers and in trade journals. Major feed companies called Lovell to ask whether they should add the vitamin to their catfish feeds, and the reply was always "yes." That year, major companies began to produce complete catfish feeds, with added vitamin C. After that, all catfish feeds sold contained what was advertised on packages as "the Auburn Formula" and included 200 milligrams of vitamin C per kilogram of feed according to Lovell's recommendations. Later, stabilized forms of vitamin C were developed, and the researchers demonstrated that these products were not damaged by extrusion and prevented broken back syndrome just as effectively.

The fish nutrition program at Auburn, the first of its kind with warmwater fish, began officially after Lovell's success in determining the cause of broken back syndrome. Research focused primarily on the channel catfish because it was such an important fish in the South. Guidelines for formulating feeds were developed, and feeding recommendations—how much to feed, how to feed in winter—were established.

~

The normal hematocrit, or red blood cell count, for catfish is not very different from that of healthy humans, which is usually 35 or 40 percent or higher.[4] A healthy count for fish is 25 to 30 percent or a little higher. Just like humans, fish are considered anemic and in trouble if red blood cell levels go much below that. John Plumb and Project researchers were therefore nonplussed when, in 1983, farmers began bringing in sickly fish that showed virtually no red blood cells at all. The fish's hematocrits registered at 10 percent or less. The researchers found only clear fluid where red blood should have been. The fish had no color. The gills were almost white—not the healthy red they should have been. And strangely, not all of a farmer's fish would be affected, only some. The same was happening in the ponds. Two or three had anemic fish in them, while others had none. The researchers were stumped. For lack of an existing name, they dubbed the condition "no-blood disease."

At that particular time, Ted Butterworth, a doctor at the University of Alabama's medical school in Birmingham, was on sabbatical, visiting Auburn one day a week to learn about disease problems in the catfish industry. Butterworth had been an army doctor in Puerto Rico and had seen many different diseases among the people there.

"You know, this whole thing reminds me of tropical sprue," he remarked when he heard about the problem. "That's a human disease caused by an increase in pteroic acid, a derivative of folic acid breakdown." Tropical sprue, he explained, occurred on Caribbean islands.

One of the cage farmers whose fish were developing no-blood disease was using a feed he had bought at a bargain price. It was the previous year's feed. After seeing eight or ten cases raised on the same feed, the researchers were almost able to predict the problem without even examining the fish.

"You're using such-and-such feed, aren't you?" they would say, to the surprise of the farmers, who would ask in amazement, "How did you know that?"

Aided by Butterworth's knowledge and experience, Plumb and Rogers began by looking into the feeds the farmers were using. They discovered that many feeds on the market tested deficient in folic acid, even though they had contained adequate folic acid when manufactured. The diseased fish had high levels of pteroic acid, which could be toxic. The researchers sampled feeds used by catfish farmers around the state, and almost all the feeds given to the diseased fish were low in folic acid and high in pteroic acid.

It was time for epidemiological studies to determine the cause. The researchers set up cages and fed the old feed to some and nutritionally complete, fresh feed to others. Within a week, the fish that were getting the old feed started dying, and diagnosis showed the no-blood condition. The experiments proved indisputably that the disease was related in some way to the old feed.

Substantial progress was made when the researchers isolated a species of the *Pseudomonas* bacterium from the fish. They then found the very same bacterium in the old feeds. This particular organism could, by itself, break down folic acid to pteroic acid. The biological culprit had been exposed. Many of the tainted feeds had come from one particular manufacturer, and a visit to the factory turned up highly unsanitary conditions in the manufacturing process. The suspect *Pseudomonas* showed up throughout the system. In addition to the entire processing line's being tainted, much of the feed had been stored for long periods.

With an effective explanation for no-blood disease, the handling and storage of feeds improved greatly. Occasionally, a case still pops up, but the mortality it causes among fish is very low and not considered significant.

In the mid-1980s John Hawke was a graduate student working on his master's degree research in the Parasite and Disease Project at Auburn. One day, a sick catfish was brought in from Georgia. According to the farmer's report, the first signs of a problem were small, white-colored areas in its skin. These had turned into open red ulcers that were now hemorrhaging severely. Hawke set about trying to isolate and identify the cause of the illness. Not long after, the disease turned up in Alabama fish and then throughout the industry.

As with channel catfish virus, researchers suspect now that the bacterium Hawke eventually found had been around a while before it started causing so much damage. When the first fish were brought to the lab, researchers had no idea what the disease was. In the course of their investigation, they reviewed slides taken of fish they had seen a decade earlier and were able to recognize the same signs. Something that happened in the industry's development during the previous ten or fifteen years had apparently made the mid-1980s a perfect time for the organism to become much more destructive. A host of changes in industry practices could have prompted the flare-up, among them environmental conditions, increased stocking densities, or even transport.[5] The disease, enteric septicemia of catfish, has since become the greatest fish health problem of the industry.

At the same time regulations began eliminating chemicals and antibiotics from farmers' repertoire of treatments, simple oxygen meters and paddle wheels were making it possible to control and counteract the effects of excess nutrients and wastes. Farmers were jumping at the opportunity to increase production. Whereas they originally stocked their ponds to produce 2,000 or 3,000 one-pound fish per acre, many began growing 5,000, 6,000, even 9,000 or more. The farmers did not believe in just stocking a few fish, according to Bill Rogers. "They figured as long as they had some water that would hold another fish, they'd put them in there. We'd have people put 25,000, 50,000 fish per acre in there and try to raise them out. One fellow had over 100,000 fish per acre."

The paddle wheel was not the panacea catfish farmers thought, however. They began to come up against the hard facts of the ecological concept of "carrying capacity."[6] Many had not listened to warnings issued by SCS professionals and scientists and suffered huge losses in their ponds. "There were people at one time feeding up to 350 pounds per acre per day. They'd put enough paddle wheel aerators in there to keep the fish alive," recalls Claude Boyd. "We kept telling them to go to a more moderate level of production and [they would] not have so many problems. The research showed that that was what was right. But they had to convince themselves. They didn't believe the research at first, and they would argue with it." It took cold experience for many to realize that there was a limit to population growth in any environment, natural or cultured, and that maintaining water quality involved more than the level of dissolved oxygen. The lesson of water quality was a difficult concept to fully comprehend, according to Elvin Wright. "Eventually, enough people looked into the issue and read anything they could obtain to try to understand it better. If you can maintain your water quality, then you can increase your stocking, your feeding. But if you don't, if you're not familiar with water quality, you better stay at a recommended stocking and feeding rate."

Most growers had not realized that the higher densities of fish in their ponds opened the door to bigger disease problems. An infectious organism that might previously have taken a limited toll in a pond could spread like wildfire under overcrowded conditions. They had at first little understanding of what was going on under the water's surface, and this was an entirely new way of thinking for them. Gradually, most settled down to more reasonable stocking and feeding rates.

While some diseases such as brown blood, proliferative gill, and winter kill persisted, producers were gradually getting answers they needed in order to keep a healthier balance in their ponds. Those who did not accept the information and advice resulting from research found themselves facing a brick wall after a few years, suddenly unable to maintain fish populations in their ponds. One important thing farmers learned was that water quality, and not just oxygen content, has a direct correlation to disease problems, according to David Pearce, who ran the gamut of skepticism, overproduction, and failures with his ponds but managed to come out successfully at the other end.

Notes

1. A raceway is a rectangular tank for raising fish at high densities in which water is constantly flowing through the system, most often by gravity.

2. The fluorescent antibody technique can identify whether a particular disease organism is present in an affected tissue sample. It works by mixing that sample with antibodies against that organism. The antibody is attached to a molecule that fluoresces under ultraviolet light, and if it matches the organism present in the sample, fluorescence occurs. Immunoassay techniques use specific antibodies that can show the presence or quantity of a particular disease-causing organism in the tissue sample.

3. Bacteriological assays are laboratory tests that indicate the presence of a particular disease-causing bacterium and the extent of its presence.

4. A hematocrit of 35 to 40 percent means that the red blood cells make up between 35 and 40 percent of the total blood volume.

5. John A. Plumb, interview by author, tape recording, 23 March 2000, Special Collections and Archives, Auburn University, Auburn, Ala.

6. Carrying capacity can be generally understood to be the number of living things that can inhabit a given area without causing damage to the environment in which they live.

13
A New Way to Raise Catfish

Oh, we had all the big dreams. . . .

—W. S. (Shep) Phillips

The four fisheries graduate students went out to the War Eagle Supper Club in Auburn one March evening in 1967 as they often did, looking forward to an evening of spirited discussion, laughter, and beer. Norris Jeffrey and Rudy Schmittou were doctoral students, and Hugh Swingle and Tom Scott were earning master's degrees. Their long debates were typically about political issues, and each of them would choose a side and argue.

The students had recently seen slides in Homer Swingle's classes, slides taken during his recent travels in Southeast Asia. Some slides had shown cages underneath houses built over a river. The cages were made of wooden slats and hung suspended in the water. They contained Thailand catfish. The residents fed the fish with cooked rice, scraps from the table, and whatever else was available. The practice had been going on since at least the late 1800s, if not earlier.

That particular evening, Scott reignited an ongoing debate on the efficiency of raising fish in cages. He argued for the practice, and Swingle and Jeffrey took the opposite position. Schmittou joined in with Scott just to even up the sides. Eventually, Scott proposed a wager:

"I bet I can raise more fish, more efficiently, in cages suspended in ponds than you can raise in fifty-five-gallon drums with a flow-through water system," he challenged.

"Okay, then prove it!" countered Jeffrey and Swingle. "We'll raise fish in barrels, and you raise them in cages, and we'll see who's right!" They sealed the pact with a beer.

During the next few days, Scott and Schmittou began designing an experiment to prove the hypothetical points they had tried to make that night. Scott was the leader, the "idea man," and Schmittou helped, just to win the argument. They worked on the project on their own, with Homer Swingle's encouragement. There was no base of previous work from which they could operate. The cage systems used by farmers in Southeast Asia had been developed and handed down within families informally and not very scientifically.

With money Swingle finagled from department funding, Scott and Schmittou searched local hardware stores for materials. They needed something strong and durable yet open enough to allow water exchange with the pond environment. They finally chose hardware cloth and nailed it to a wood frame, creating cages one cubic yard in size. They had to add weights, because the float they attached made them too buoyant. They made changes to the project as they went along, keeping the ideas in their heads and never writing anything down.

As it turned out, the attempt to raise fish in barrels failed. The idea was not a bad one, but the system available at the time to get flowing water was unreliable. When the system was working properly, the fish did very well, but when it was not, they had problems.

The information that came out of the cage experiment surprised everyone. Scott and Schmittou had stocked the fish at varying densities between 25 and 175 fish per cage. Knowing that density would be a limiting factor, they had guessed that growth to an average of one pound per fish (100 pounds per cage) would occur best at a density of about 100 catfish per cage. The results at the end of the trial were opposite to their expectations. The cages with the best production were the ones stocked with 175 fish, followed in order by those stocked with 125, 75 and 25, respectively. The fish had grown very well at high densities. In fact, the higher the density of fish in a cage, the better the production. As it turned out, territoriality was an essential factor in making that possible. As long as the fish had space in which to show territorial behavior, apparently none fared very well. Crowding the cage with more fish dampened that behavior, and all had a more equal chance. At the time, however, Scott and Schmittou knew only that the cages not only worked, they showed great promise.

Tom Scott graduated in 1967 with his master's degree and left Auburn. Schmittou, who still had work to do on his doctorate, had become interested in cage production. Both Homer Swingle and Wayne Shell encouraged him to go on with the investigation. That year, Schmittou began the first scientific research on cages. This time, he was trying not just to win a bet but to refine the ideas developed in his early experiment. Nearly every type of environment in the ponds on the experiment station land north of town became host to cages he built. He began stocking them with channel catfish and eventually expanded to eighteen different species of fish.

Two primary problems appeared early in the trials. After about a year, the wire screening began to corrode where it was stapled onto the wood frames. Second, the wire posed problems with stocking and with maintaining the health of the fish. Small fingerlings slipped out through the mesh. Larger ones, wild as they were, tried to escape, ran into the wire sides, and injured their mouths. Bacterial disease resulted. Eventually, Schmittou found that conditioning the fingerlings first in a smaller area enabled them to recognize and accept the cage's limits.

A great deal about the dynamics—the biology, chemistry, and physics—of what was going on in the cages remained unknown. Obviously, much could be learned about this apparently promising method of raising fish. At the end of the second year, Schmittou changed his dissertation topic to caged fish production. It was the first known research ever done on cages and certainly the first in the United States. The cages were entirely different in design and system of production from those used in Asia because the system was feed-based: the fish ate a nutritionally complete feed rather than scraps. In addition, the cages were designed to take advantage of water exchange in ponds whose quality was constantly monitored.

~

Many discussions accompanied the research—discussions with Swingle and among Schmittou and his friends at the War Eagle Supper Club. Cage production had clear advantages. One could separate fish species. One could harvest whenever one wanted because the fish were right there, on hand. Recapturing the fish when they were loose in the pond was much more difficult.

Beyond these advantages, however, and beyond the curiosity, very few people thought cage production had much application. Most thought the cages were impractical. Why use a cage when you can stock the entire pond and get more production? Some people criticized the idea strongly. Regulations prohibited putting cages in reservoirs or in open waters in most situations, such as in the TVA reservoirs. Concerns about environmental conditions were on the rise, and critics thought water quality would be affected. Also, cages would affect aesthetics and possibly the movements of boats in public waters. Some countries without adequate food resources might do it, but in the United States there was generally no such need. Late in 1970 several FWS researchers condemned cage production, maintaining they caused disease, would never be efficient, and were a waste of money.[1]

Swingle and Schmittou talked a long time about it. Swingle just shook his head and related the difficulties he had experienced getting research on ponds started more than thirty years earlier. "People don't understand," he consoled his student. "If it's totally different from what they're accustomed to, they will reject it, and it will take time." He continued to find one way or another to pay the costs of Schmittou's research. "It will come, so don't worry about it," he encouraged.

In 1969 Schmittou completed his dissertation, reported his findings at a professional meeting, and left Auburn. Besides a few master's theses involving cage-raised fish, work on cage production essentially disappeared. He returned to Auburn around 1980 as a member of the faculty and with students began using cages in the master of aquaculture program to demonstrate production principles or study problems like diseases. The fish were much easier to observe closely in cages than in ponds, and students were able to see the fish and handle them. They could see results within three or four weeks instead of waiting until harvest time. Cages were definitely an advantage in teaching.

The hilly landscape of the Piedmont region of northeast Alabama provides plenty of hollows into which water drains naturally. In some areas, the soil has enough clay to hold the water well. The growing season is a little shorter than in Alabama's Black Belt, but water quality is basically good. In the 1980s there were a fair number of Piedmont ponds in which farmers had been raising catfish for years. Some of the farmers had made good money from the fish as early as the 1960s, selling them for as much as a dollar per pound. Pond fish production there was not widespread, however, and those people who had ponds often had problems harvesting them. The steep, irregular topography and the stumps and brush left on the bottoms made seining impossible. The corrugated metal pipes used to drain the ponds had begun to corrode, and some could no longer be emptied for harvesting.

The SCS had for some time been administering a program in rural counties to conserve and develop natural resources for economic use. Volunteer Resource Conservation and Development (RC&D) Councils organized as part of the program were working together with county commissions, supervisors, and other groups looking for ways to build up their local economies by creating supplemental or alternative income for farm owners. The

Coosa Valley RC&D Council hoped to find something that would help farmers in the Piedmont. Dairy farming, row crops, and occasional timber harvesting provided limited income, and residents knew of relatively few other possibilities. Many farmers had taken jobs in surrounding towns while still partially working the land.

No one quite remembers how it came up, but late in 1984 or early in 1985, the idea of possibly developing a small catfish industry came up in conversations in the office of Robert Berry, SCS area conservationist in Oxford. SCS biologist Dave Kelly and Ron Burdette, who coordinated projects for the Coosa Valley Council, both thought the idea had promise. Before they could look into the possibility, however, they needed to know if there were even enough ponds in the Piedmont to make such a project worthwhile. After conducting a survey of pond owners in eight different counties, they learned that more than seventeen thousand acres of ponds existed already, plenty enough to give fish farming a try.

The men were aware that there were important differences between the Piedmont and West Alabama. Clearly, the way catfish were raised in the west was not directly applicable in northeast Alabama. The hilly land set certain constraints that were no longer limitations in the Black Belt, where ponds of more even depth now predominated. Managing relatively isolated pond sites for commercial production would not be economically viable. Hill ponds did not lend themselves to harvesting by seine. The men could see that pond production in the Piedmont would not be able to compete with the existing industry.

Kelly remembered hearing in the past about raising fish in cages. He knew Rudy Schmittou from years past when they had both worked as fisheries biologists for the state. He also knew how to bring people together to solve problems. Meeting with Schmittou was a logical next step.

Schmittou was out at the experiment station ponds with his students working with cages when Kelly, Burdette, and Berry drove up. After introductions, the men stood on the bank and talked as the students worked. Kelly asked Schmittou to work with them to see whether caged catfish production might not be feasible in Piedmont ponds. Schmittou, always interested in helping people and fisheries development, agreed. Wayne Shell, who headed Auburn University's fisheries department, was always ready to support fisheries extension efforts and consented to Schmittou's spending time on the project.

Schmittou put considerable effort into getting the project off the ground. Under his guidance, master's students helped the SCS survey the land for possible pond sites, locate water sources, find ways to direct the water, and lay out plans for individual ponds. The potential they uncovered was surprising: close to 165,000 acres could feasiblely be turned into ponds. Kelly and Burdette talked to Shep Phillips, a farmer who was also a Coosa County supervisor and on the RC&D Council at the time. In April 1985 Phillips proposed the idea to the Board of Supervisors and the council. The board and the council found the project promising and worth undertaking. They handed the matter over to Burdette to coordinate, and the project was under way.

In contrast to the way the West Alabama industry got its start, the cage industry got off the ground through active cooperative effort from day one among the SCS, the farmers, and Auburn University. The Piedmont project did not develop in its early days as a "private" enterprise in the usual sense of the term. Decisions were generally made by a group and guided by the input and expertise of public employees.

The last thing the organizers wanted was to encourage farmers to invest time and money in ponds, cages, and fish and then find themselves with no market or some other major problem. A trial demonstration seemed the best way to test the whole idea. Thus evolved the "eleven-acre" demonstration project. "We did not work to get a lot of people the first year," Shep Phillips remembers. "We wanted to keep it small, work the bugs out. We didn't want to promote something through the council that would cost a lot of people a lot of money. It was a new industry, and we didn't want to get a lot of people to think they were going to get rich."

Burdette and representatives from the RC&D Council began to meet and plan with Shell, Schmittou, and Kelly. The council put up some of its funds for the project, and the SCS offered personnel to assist. The plan was to recruit volunteers who would put a cage or two in their ponds and stock and manage them. The volunteers would receive training and guidance from the organizers. Burdette, with help from Schmittou and the others, would monitor production progress and results. A timetable was put together. Later, opportunities would be offered for people in the counties to view individual demonstration cages and learn more about catfish farming.

Over the remaining months of 1985 and early 1986, Schmittou, Kelly, and Burdette developed guidelines for the participating farmers. Schmittou hosted workshops and seminars showing the operation and general principles of cage culture at Auburn University and in the counties (see figs. 15 and 16). Kelly and Burdette took a group from Chambers County to Hale

Fig. 15. Net demonstration, Auburn University, 1986.
Potential producers learn about cage construction during the first year of the cage project. Rudy Schmittou (*far right, toeing the net*) is talking to farmers and fisheries graduate students at Auburn University. (Photo courtesy of USDA, Natural Resource Conservation Service.)

County in June. They looked at farms and hatcheries and toured large and small processing plants. They left feeling confident that they would be able to obtain fingerlings from a hatchery in West Alabama and that a processing plant there would buy their fish, at least initially.

District conservationists with the USDA's Natural Resources Conservation Service publicized the county workshops to draw initial volunteers. Some were participants in the original RC&D planning process who wanted to try it for themselves. Others were local farmers who had ponds or were considering building ponds. All were intrigued by the idea. "It was just real interesting. We had the water. It was different, challenging, a way to learn, a way to diversify, and maybe supplement your income," recalls Teresa Brooks, the project's only female volunteer and later a successful cage producer. Eventually, eleven volunteers were chosen from counties in the area. Spreading the trials out among the counties would promote wider visibility and communication.

Fig. 16. Cage harvest demonstration, Auburn University, 1986.
Rudy Schmittou (*to right of scale, in profile*) demonstrates a cage harvest at Auburn University for producers during the first year of the cage project. A nylon cage is being harvested using backhoe, boom, and scale. (Photo courtesy of USDA, Natural Resource Conservation Service.)

The planners wanted to supply the farmers with all the necessary components of the experiment. In order to help maintain uniformity during the first year, Shell agreed to have the Auburn fisheries department purchase the equipment and supplies and resell them to the participating farmers. Variations in cages, fingerlings, and feeds would thus not be a concern, and the volunteers would not have to invest extra effort to get what they needed. The organizers discussed fingerling sources and interviewed potential suppliers. Time was also spent on planning for a possible local hatchery and processing plant.

March 1986 came, and, ready or not, it was time for stocking. Cages specially ordered from Mississippi arrived late and did not meet the size specifications, but the farmers used them anyway. The cages had to be placed in deeper water in order to allow space for the fish waste to sink away from them and to keep the fish off the bottom of the pond. Most of the farmers already had piers jutting out into their ponds to which the cages could be attached and from which they planned to feed the fish. Others ran cables between trees across their ponds to which they would tie the cages. For them, feeding would require going out in a boat. The fingerlings that first year came from a hatchery in West Alabama, hauled in one of Auburn University's fisheries tank trucks. Eleven cages in five different counties (Clay, Randolph, Coosa, Chambers, and Tallapoosa) were each stocked early in March with around twenty-six hundred young catfish.

Almost right away, the project ran into near disaster. The team administering the project stood watch over the cages through the first night after stocking. Within the first twenty-four hours, many of the fingerlings died. The team realized only later that the fish had been overly stressed by being hauled seven or eight hundred miles and then dumped into a situation to which they were not accustomed. Fortunately, enough survived to allow the demonstration to continue.

The trial demonstration was a learning experience for everyone. Once the fingerlings were established in the cages, the farmers settled into a daily routine of checking and feeding. Infectious diseases did not seem to be a problem at first, but not everything went smoothly, especially at feeding time.

In nature, channel catfish live on dark river bottoms and in big rivers, where fluctuations in oxygen and temperature are relatively minor, if they occur at all. They are still basically wild animals and not adapted to being enclosed in a small area near the surface, where they are less protected from the sun, predatory birds, and the more extreme changes in light and temperature that exist in ponds.[2] "He feeds at night, and you're bringing him up here in a suspended cage, and you're feeding him a prepared feed. Every time you'd go out there, you'd be scaring him to death. We even found that just your steps on that pier sometimes would make that animal hesitant to feed."[3]

The cages had a small floating ring at the surface to contain the feed the farmers poured in. Not all were able to feed well because the more aggressive ones took the largest portions and the less aggressive ones got only what was left. As the months went by, the fish grew to widely varying sizes. They needed to grow to more-or-less uniform size if the farmers were going to be able to sell them.

Buying feed in bulk was considerably cheaper than having the farmers buy their own, but providing the feed was a headache at times. Burdette and others would buy several tons at a mill and truck it to one farmer's property to distribute from there. Not uncommonly, some volunteers picked up their share, while others did not. Burdette was often left sitting for hours in the truck, waiting for a farmer to come get his feed.

Schmittou supervised the monitoring of cages. He shared his observations with Kelly, Burdette, and his students, and taught them how to assess the fishs' progress. Schmittou knew that the farmers would have a tendency to overfeed, so he kept an eye on feeding rates and advised changes whenever necessary. The farmers looked forward to the regular visits when people would come to talk and to weigh the fish.

The monitors also checked water quality, sometimes sampling ponds for an entire day. Schmittou and his students spent many a night during the warmer months bent over cages with flashlights, oxygen meters, and notepads. Midnight and two o'clock in the morning were not the safest times to go traipsing around people's private property, they learned, as Schmittou on one occasion nearly found himself the target of a worried landowner's gun.

Together, the team worked to inform the public and potential cage farmers. Burdette and Kelly toured the demonstration sites with groups of people who had come from near and far to see how this curious way of raising fish was working. During June and July alone, tours brought around two hundred people face-to-face with cages in the counties. A June field day drew farmers from all the counties participating in cage culture to AU. That same month, Schmittou, Kelly, and Burdette introduced television audiences to the fledgling project. June also brought two thousand dollars from the state farmers' organization and fifteen thousand from the state extension system to be used for one year of research on the project.

~

Everyone's optimism was seriously tested by major problems before the growing season ended. Weatherwise, the group could not have chosen a worse year for the trials. By early August, it was clear that the fish were not growing as expected. Feed was going uneaten. No one could have foreseen that 1986 would bring one of the worst droughts Piedmont farmers had ever known. Weeks and months dragged by without rain. Water temperatures climbed into the nineties, and water levels dropped relentlessly. Finally, rains relieved the situation, but those months were worrisome for everyone. In retrospect, one had to agree that the weather provided a good test of whether the project would work before more people began to invest in cages.

As summer drew to a close, the issue of where the farmers could sell the fish became urgent. Having come up with no firm commitments from processors, the organizers were at a loss. West Alabama plants were unwilling to travel so far for so few fish. The outlook was generally dim. Farmers sold a few to friends and neighbors. Ron Burdette prepared to help farmers take fish to Birmingham and other cities to sell live from tanks on downtown squares or at major crossroads. Finally, Wayne Shell brought his good will and Auburn's resources to the rescue once again and offered to buy the fish, at the going rate, to sell at Auburn's fish market next to the university's own fishponds.

The trial demonstrations accomplished that first year what they had been set up to do. The cooperating farmers had learned a great deal about how to raise catfish successfully in cages. Most, if not all, decided to continue, and others were interested in following suit. Rudy Schmittou was receiving many inquiries from people interested in the cages.

Shep Phillips thought it would be good to have a local source for cages and decided to build cages himself. He found a netting material that would work better than the nylon of the first cages and began to experiment with ways to build them. Bouncing ideas off of Rudy Schmittou and others, he settled on a design and put together eleven cages, one for each of the trial volunteers. The project used Phillips's cages in its second year. The polypropylene half-inch netting he used was similar to seine material and sewn together into a box shape that hung from a rectangle of two-inch PVC pipe about twelve feet long and six feet wide at the top. Weights hung from the bottom held it down about forty inches in the water.

The organizers evaluated the project's achievements in 1986. Schmittou examined the experience from the scientific aspect and Burdette from the practical one. As usual, Kelly helped with whatever needed doing. Many of the problems were fairly simple to correct, and enough information had been gained to write a manual on raising channel catfish in cages suspended in ponds. Obstacles still needed to be overcome, and developing realistic strategies for selling the fish had top priority.

To avoid another fingerling dieout at stocking time, a nearby fingerling source had to be developed. During the 1980s the U.S. Department of Agriculture (USDA) was in the process of ending subsidies to dairy farmers, and, like many other farmers at the time, Elbert Butler was thinking of closing down his dairy cattle operation in Sylacauga. Several years of dry weather had persuaded him to stop growing soybeans and corn. By 1987 he was looking for some other type of farming enterprise. He spoke one day with Claude Friday of the SCS about his idea of converting some swampy land on his farm into a large lake. Friday's enthusiastic response, "Say, that'd be great for the cage industry!" prompted Butler's interest in the catfish business.

Butler's decision to establish a hatchery was nurtured in talks with Friday, Schmittou, Kelly, Burdette, and others. He traveled to West Alabama to look at operations there and found himself drawn increasingly toward the idea. Once he decided to do it, the cage project organizers jointly planned ponds, tanks, water systems, and a building. The entire setup was designed by the group to fit Butler's situation, all with an eye to keeping his costs down. The hatchery was in operation by February 1988, in time for stocking in late spring and summer of the project's second year. It served Piedmont cage producers well.

Expectations had been that if cages were stocked with about 2,600 six-inch fingerlings in the spring, the fish should grow to one-pound size and make a very saleable crop. Most of the farmers had ended up the first year with only about 1,800 pounds of grown fish, however, and many of the fingerlings had never grown out. These were known as "shorts." With time, a good balance was found, and the stocking rate was cut in following years, first to 2,000, then to 1,800, and finally, around 1990, to between 1,500 and 1,700 per cage. At that level, the farmers ended up with two to three thousand pounds of fish and no "shorts" (see fig. 17).

Planning and managing the developing cage industry continued in a unique fashion, just as it had begun. Interdependence and cooperation between all the contributors was a key element. Rather than working each from their own specialized approaches, the organizers worked together to deal with all the problems and needs of the farmers. If one of them slipped up, the entire project could fail.

~

Feed posed a problem from the beginning. Cage-raised fish have somewhat different nutritional needs than those raised loose in ponds. They are more susceptible to stress from a variety of sources, including nighttime drops in oxygen levels. The constant checking to track their growth during the first year—the dipping up, weighing, and counting—caused additional stress. The stress appeared to result in higher susceptibility to parasitism and disease, which then had to be treated with chemicals, and the use of chemicals itself stressed the fish's systems.[4]

Unable to range throughout the pond to find a variety of natural foods

Fig. 17. Demonstration harvest from a cage in Chambers County, 1986.
A demonstration of an actual cage harvest at a grower's pond in Chambers County. The cage yielded twenty-six hundred pounds of catfish at 152 days after stocking. *Left to right*: Rudy Schmittou, Ron Burdette, and an unidentified student. (Photo courtesy of USDA, Natural Resource Conservation Service.)

to supplement their diet, caged fish were much more dependent on nutrients provided in the feed. The best way known at the time to build up the fishes' resistance to disease was to increase the vitamin and mineral content of the feeds, in particular vitamin C. By now, the feeds used in West Alabama had everything that was needed for the fish loose in ponds. Yet the cage-raised fish seemed to be under greater stress and to need even more of the vitamin. Schmittou suggested that the vitamin C content be increased, and Burdette and Kelly pleaded during the first few years with a number of manufacturers for a feed with higher vitamin C content specifically for cage culture. They encountered resistance because the demand was much lower than in West Alabama or Mississippi.

The two men were finally able to convince researchers at several companies of the merits of their arguments. BAMA Feeds was the first manufacturer that, on the basis of a letter from Tom Lovell stating the specific types and quantities of nutrients for cage culture, began to make the feed commercially available around 1988. With Wayne Shell's approval, Schmittou was able to work out a way to sell the farmers fresh feed through the Auburn fisheries department, partially solving the freshness problem they had with the cheaper feeds obtained in stores. After the first year, when the participants knew better what to expect, the fish were sampled less often. With less frequent sampling and the eventual use of high vitamin C feeds, production went up, and the incidence of commonly seen diseases dropped dramatically.

Cage farmers were well aware of the advantages of caged fish production. The big question was where to sell the fish. The second year saw more pounds of fish produced than the first. Some of the farmers used more cages. Burdette, Kelly, and others had told them from the start that no markets had been established yet. Farmers had been advised to book sales for the fish among neighbors and friends. People in local communities bought 10 or 20 pounds for themselves and their neighbors for neighborhood fish fries. A few times a year, church groups bought 50, 100, or 200 pounds, perhaps, for fish fries. Twenty pounds here and 40 pounds there hardly took care of the 2,000 pounds that one cage alone produced, much less two or three of them.

By the 1980s many reservoirs and other water resources had become available to the public for recreational fishing. The fee-fishing ponds had little business and provided only a limited market to cage farmers. Fee-fishing pond owners could pick up a tank truck of cage-raised fish for 80 cents a pound one day and transport them to their ponds. They could then let people fish them out the very next day for $1.25 a pound. Little satisfaction existed for cage farmers in knowing that, after investing six months' worth of feed, materials, and daily time and effort, fee-fishing pond owners could make at least as much or more profit from the fish in just one or two days.

Attempts to promote catfish in local and nearby chain restaurants resulted in a few sales of limited quantities. However, the group was burned at least once when restaurant owners committed to buy fish and then failed to come through. The farmers desperately needed a stable market.

The number of producers increased each year. In 1988, the third year of the project, Red Kennedy, a pond owner in Clay County, was one of the newer ones. He had bought five cages in anticipation of raising and selling a lot of fish, but the more he thought about it, the more he realized how difficult it was going to be to find a buyer for five tons of catfish. He and his wife came up with a solution: They would build their own catfish restaurant. And they did—on their farm, on a dirt road, in the middle of rural Clay County.

Friends exhorted, "Red, folks are not going to come way out here to eat fish!"

Red responded with a grin, "Well, that's what everyone says. But if they don't, we're sure going to have a nice house to live in."

So Red began to serve nothing but cage-grown catfish out in the "wilds" of Clay County, Alabama. His first customers came on July 19, 1988. They were the RC&D Council, coming for a lunch meeting. Everyone agreed the fish were delicious. Caged fish always tasted good, and Red sure knew how to cook them. Word spread, and others began to come. Red's business grew steadily.

Before long, Red needed more fish than his own cages provided. Encouraged by the unexpected success, he turned to other Piedmont producers. He contracted to buy their fish and built a small processing plant in his old milking barn. He and his one processing employee drove a tank truck around to farms once a week, picked fish up, processed them, and served them Thursday through Sunday. "This was a big milestone for the project," Burdette recalls. "Red was going to use all of the fish we could produce. . . . He was going to pay seventy-five to eighty cents a pound for them, which was good. In the next short period of time, he was running around seven thousand pounds a week through that little restaurant. These were kind of glory days of the industry, so to speak, as it really began to blossom."

The farmers were trying to produce as many fish as Red needed. True, it was a relatively small amount compared to what was raised and sold in West Alabama, but they had gone into cage production from the start only as a side occupation or hobby. The demand was exactly what they had been hoping for.

Compared with early expectations, things looked as though they were going swimmingly. Farmers were getting feed conversions averaging 1.2 or 1.3, meaning the fish gained close to a pound of growth for every 1.2 or 1.3 pounds of feed they were eating. In an industry in which anything below a rate of 1.5 was considered excellent, these were exciting numbers. According to calculations, about forty-one cents were being invested per pound of fish harvested. That number included cost of nets, feed, and fingerlings. The fish were being sold for close to twice that amount! What was left after sales looked like good money to the farmers. Within a few years, Red's restaurant was the primary market for Piedmont cage producers.

The cage project continued to grow. By fall of 1989 more people were raising catfish. About 130 acres of ponds were devoted to the project. Red's restaurant was taking close to 250,000 pounds of fish a year, but the farmers' nets were giving a total crop of 400,000 to 500,000 pounds more than that. They had to find other places to sell their fish, but many were reluctant to look for markets on their own. Like many farmers, they generally enjoyed and excelled at producing but neither enjoyed nor felt adept at marketing.

With time and with growth came other needs. Rudy Schmittou had left to work overseas in 1988. Dave Kelly had had to shift his attention to the state's wetlands. The farmers no longer had the help of a fisheries specialist available to them. Although cage production had become well established in the short time since the demonstration project, the farmers still needed assistance on a daily basis from someone who could work with them on marketing, diseases, stocking issues, and other problems.

The logical approach to these problems seemed to be to form a producers' association. With fewer than two hundred acres of ponds, the industry was still a small one. The pond sites were scattered throughout a number of counties. The farmers reasoned that strength in numbers as an association would help them obtain not only an aquaculturist but also economic clout with feed manufacturers and more markets for their fish. By April 1990 the Piedmont Association of Caged Fish Producers was in place with papers and officers.

A group of county supervisors traveled to Montgomery to draw state legislators' interest in the developing industry. The Alabama Department of Economic Affairs granted the association twenty-five thousand dollars. The RC&D Council contributed a little over eighteen thousand. With this, the farmers could hire their first cage aquaculturist. The money had to be requested anew every year, so the position was not permanent, but in April 1990, fresh out of the Auburn fisheries master's degree program, Claude Reeves began the job.

Reeves threw himself immediately into his primary assignment with Ron Burdette—finding places for the farmers to sell fish. The total cage harvest was not big enough to supply large markets. They sought small niche ones and worked with the growers to help transport fish. They contacted every potential buyer they could think of, including small restaurants and processors in Alabama and west Georgia and livehaulers. They had limited success, but as a whole, things still looked promising.

After less than a year, Reeves found a permanent job and handed in his resignation. The association found itself looking for a replacement. Because of the uncertainty of funding, the growers had to find a way to finance a permanent aquaculturist's salary. Only a more established, institutionalized route of funding would enable them to do this. The farmers had obtained an annual appropriation from the state. Burdette, Kelly, and council and association members appealed to Wayne Shell to take the money and hire a permanent aquaculturist with it through established channels at AU. Arrangements were made for money to go through Auburn's fisheries department to provide for the position. Finally, the cage industry could have a permanent, full-time extension aquaculturist in the Piedmont.

A lot of the things we dreamed of never came true.

—Shep Phillips

By the early 1990s Red Kennedy's restaurant had built up a fairly steady clientele, and business was good. The farmers had had some success at finding other, small niche markets and thought the arrangement with Red would continue for a long time. Red's was a major market for the growers' fish, but already in 1990, cage production was outstripping what he was able to buy. The farmers' hope did not materialize.

While they knew they could not compete with pond producers and the larger, West Alabama industry, cage producers were still subject to the overriding economics of the industry as a whole. In the early 1990s the price of pond catfish dropped dramatically. Large-scale processors were buying them for close to fifty cents a pound, about what it was costing cage farmers to raise the fish. At that price, the plants could buy, process, and sell pond-raised fish significantly more cheaply than cage farmers were willing to sell theirs, unprocessed. Some cage farmers who had been selling fish for a dollar or more per pound did not want to lower their prices so much that their profit would disappear.

Red Kennedy knew that buying catfish from the Piedmont farmers would not be a smart business decision. Seventy-five percent of the fish his customers wanted was filets, which he could get at less expense, ready to cook, directly from West Alabama. Why should he pay more than necessary for cage fish and, in addition, transport and dress them himself? He decided to purchase from the processors.

Red's decision came as a surprise to some of the farmers. "They had no place to go with their fish," remembers Elbert Butler. "It left them stranded. Those guys wound up with a bad taste in their mouth. They weren't goin' to stock anymore. That's when it started goin' the other way."

While it may have seemed to some at the time as though Kennedy's decision started the downturn, a closer look shows other factors figuring importantly into that shift of fortune in the cage industry. The project had started out primarily as a planned, group enterprise, yet general agreement among the farmers on a number of important issues remained consistently out of reach. Who would have predicted that the independent nature of the farmers would be any kind of stumbling block? No individuals, groups of individuals, or common concerns could apparently convince some of them of the need to unify on those issues or to invent other solutions. Possibly as a result, benefits for all that might have been obtained through concerted action were not realized. An example was the problem of getting agreement to buy fresher feed at lower prices, according to Shep Phillips. "We were going to have one central place that was handling the feed, and all of it would come there. But it's like everything else. You've got the individuals [with] their own ideas. They want Goldkist, or they want this, or 'I'm not going to buy from him because of. . . . ' That type of thing."

The distance between ponds was also a practical constraint for some. The hill ponds were situated much less densely than the levee ponds in the Black Belt, and the cages were distributed across a wide area. "Folks from Calhoun County wouldn't want to ride all the way over to Eclectic to pick up feed," says Phillips.

The farmers never agreed among themselves on a price to charge Red Kennedy. More significantly, perhaps, they never agreed as a group to try to hold or stagger their harvests in order to provide him with fish year-round. Even though the seasonality of cage harvests limited their market options, growers generally resisted lowering their price to compete with those in West Alabama. By the mid-1980s seasonality had been overcome in the pond catfish industry, where year-round stocking had become virtually universal.

A number of reasons likely contributed to the reluctance of Piedmont cage growers to raise fish year-round. They did not want to expend feed to maintain nongrowing fish through the Piedmont winters, which were generally somewhat longer and colder than those in the Black Belt. For these farmers, the cages were a sideline, a source of supplemental income. They had made much smaller investments of land and money in their operations than the farmers in West Alabama and were not as motivated to put out more for just a part-time occupation or hobby. That fact likely made arguments for year-round harvests less effective.

> We never thought we needed a large processing plant. We're looking more at small processors doing 3,000, 4,000, or 5,000 pounds a week. We've never recommended to anybody, "Go out here and build 100 acres of pond and get rich raising fish in cages." That was not the intent. It was strictly to utilize existing ponds and get a little additional income.
>
> —Ron Burdette

Perhaps some cage growers could have been convinced to stagger their crops if they had been dependent on large processors for a market. The limited size of their harvest, the distance, and the scattered distribution of cages across the Piedmont landscape made the prospect unattractive to plants in West Alabama, however. Hopes had at one time been high that perhaps other restaurants might begin to take the fish and that Red Kennedy would have been able and willing at some point to build a larger processing business in addition to his restaurant. Existence of a nearby processing plant, however small, would perhaps have provided the motivation the farmers needed.

Yet there was some question as to whether it would have been feasible for anyone to undertake such a project. Even a small processor would have been competing with people in West Alabama and Mississippi. Large processors could process millions of pounds, market their fish widely, and still make money with only a few cents of profit on each fish. That could not be done in a small-scale business. Elbert Butler, who operated a hatchery for a few years, considered the possibility of processing. Hatchery and processing businesses seemed like they might be mutually beneficial. "If I can process these fish, then I can sell more fingerlings," he thought. "As I pick up the fish, I can stock the cages." He never did find it feasible to invest the money. Even though his hatchery had operated smoothly, when cage growers began "dropping out," he closed the hatchery down.

A greater demand for the fish may actually exist than the present system can supply. The growers' association is still being approached by prospective buyers. Without a processing sector associated with the industry, however, requests that come from larger buyers like chain restaurants cannot be considered.

Anything other than a very small plant would encounter certain other problems as well. Once a plant reached a level of seven thousand pounds of fish a week, the state health department had to become involved. The state Department of Environmental Management would also have to monitor wastes. Not least, the plant would have to sell what it processed, and the processor would have to put a great deal of effort into obtaining, developing, and maintaining markets. All in all, the investment appeared to be more trouble than it was worth.

There have been ebbs and flows. Many of the smaller markets, whether processors or restaurants, tend to be somewhat unstable, even more than fifteen years after the cage industry began. Farmers are able to sell their fish to a market here and another there. One operates for a year and closes, and another appears. On occasion, farmers have decided to process fish themselves. Often, they do it for a short time, then find that processing was harder than they had expected, and processing catfish without suitable facilities is even more difficult. They finally choose not to continue. For all these reasons and more, Red Kennedy's decision to buy pond-raised fish highlighted a significant problem the cage industry has faced in its development.

The cage industry developed very differently from the pond catfish industry for a host of reasons. The lay of the land and sources of water are more obvious differences. Others are less apparent but no less significant. Although commercial cage production started as a group concept and was established through deliberate group effort, differences in farmers' commitments to the group prevented much-needed consensus.

Without a processing sector present, marketing was to a great extent up to the farmers themselves, to their agent, or to a resource agency employee. The relative size of cage growers' investments affected their decisions and expectations. The few who made larger investments tended to sell many more of their fish. West Alabama farmers, who had in general devoted much larger acreages to catfish and were more dependent on the fish for their primary income, had stronger motivations to succeed financially. From the beginning, many cage producers' decisions were driven by factors other than the need for a primary income or perhaps even a significant return on investment. According to Burdette, "We're not talking about a big industry here. We're talking about alternative income and, to be honest with you, probably the therapy that people got out of this. A lot of the retired people would go down there every day. They'd see those fish. They'd get to feed them. They got a lot out of this."

For Butler, the prospect of eventually making money was enough to keep him and his wife operating their hatchery for a few years. "At that time, that was all I was doing," he notes. "Wasn't makin' any money at it, but we was doin' it and havin' a good time. We'd had a little money put back to keep us and tide us over for a while. That's what we used to live on, with the anticipation of makin' money." The Butlers' hatchery operated quite successfully, and demand for their fingerlings had grown. However, even that demand was not enough to overcome the larger economic turn of events in the early nineties, especially when combined with the debt they had incurred. "I borrowed in the beginning," Butler recalls. "I borrowed about fifty-five thousand dollars to build the ponds and put everything together to get this operation going. I paid the loan back on the ponds, but then it just didn't seem feasible to keep goin'."

~

Cages clearly have a positive role to play in aquaculture in the realm of education and research. As a way of growing fish populations in a contained situation, they provide an excellent teaching tool in school systems. Nationwide, cage culture provides a way for students from kindergarten through university to observe the production process and actively use what they are learning in the production and maintenance of the fish. Because conditions in cages can to a certain extent be controlled, they provide an excellent opportunity for university research into specifics of fish culture such as nutrition, treatment of disease, or genetics.

Cage production of catfish and other species was and continues to be practiced in many other states and countries. It has spread to other areas of Alabama that could potentially become as active in it as the Piedmont. Via fisheries extension from Auburn, Rudy Schmittou and others have helped transfer the techniques and knowledge to China and elsewhere in the world, where populations are in need of greater food resources. Schmittou's guidelines for cage culture, which began as the directions written in the mid-1980s for the demonstration project, have been translated into twenty languages and are used in many countries.

Around 2000 the Piedmont cage industry had as many as 150 different producers. The industry encompasses a total land area not dissimilar from that used in West Alabama. Production has stayed fairly consistently good.

Producers stand by their claim that cage-grown fish have a better flavor than pond-grown ones. The ponds are not overstocked, and therefore water quality tends not to be an issue. The main problem appears to lie in the lack of adequate and consistent markets.

The cage industry is not as stable as in West Alabama, but farmers are finding they can produce and make some money raising caged fish. Small niche markets continue to open up. They may close, but others appear. The feeling is generally that producers could take advantage of the bigger market potential that exists if a processor would choose to go into the area and the producers choose to commit to providing the processor year-round. In the meantime, a small processing business and fee-fishing operations take nearly all the fish presently available, and the remainder is spread out among numerous restaurants. Through the efforts of the Piedmont Association of Caged Fish Producers, the growers have irons in the fire for other, new marketing possibilities. Their story is by no means over.

It's at the point now, that it can get as big as they want it to get.

—Ron Burdette

Notes

1. Homer N. (Rudy) Schmittou, interview by author, tape recording, 13 April 1998, Special Collections and Archives, Auburn University, Auburn, Ala.
2. Schmittou, telephone conversation with author, 16 March 2004.
3. Ron R. Burdette, interview by author, tape recording, 6 April 1998, Special Collections and Archives, Auburn University, Auburn, Ala.
4. Schmittou, telephone conversation.

14
Production Matures

> The industry is showing strong evidence of stabilization and maturity. The era of the hobby farmer is over, and the professional farmers are in full command.
>
> —*Commercial Fish Farmer*

The commercial catfish industry went through in forty-plus years what has taken a century or two, if not more, to unfold in other, longer-established agricultural industries. Its development was not a steady one, and the first two decades in particular were unsettled. Waves of successful years lured throngs of hopeful producers-to-be into the business, and the amount of pond acreage rose. Processing plants and catfish restaurants sprouted up and flourished, and fee-fishing operations proliferated, especially near population centers. Each time, increased production flooded the market with fish. When the value of fish would then drop, many fish farming enterprises and related businesses went bottom up, not just in Alabama.

Various factors caused difficulty for the Alabama industry during downward trends. Wild catfish, caught in rivers of Central and South America, competed with farm-grown channel catfish in markets and restaurants as early as the 1970s. The imports filled a significant portion of the demand for food fish. Periodically, unusually high numbers of fish produced by Mississippi's larger industry created gluts in the market. At one point in the seventies, the catch of anchoveta from coastal Peruvian waters dropped dramatically and failed to provide enough of the small fish that were used in fish feeds. Fish meal prices skyrocketed as a result. Farmers could not pay for feeds and had to raise fewer catfish. Many Alabama farmers who relied mainly on rainfall to fill their ponds were especially hurt by droughts in the 1970s.

By 1980 the number of fish farmers in the state was nearly half of what it had been ten years before. So many people lost their livelihoods or their land that observers shook their heads in dismay at the "regrettable" and "very terrible" trends. Economic analysts used euphemistic terms like "shakedown" and "consolidation" periods to describe the paring down of the industry to include only those able somehow to "hang on."

Some catfish farmers managed to stay in business. How were they different from the ones who dropped out? The size of farms appeared to be one factor. Early on, most fish being sold in Alabama had been raised by farmers with small pond acreages. But small-scale growers had a harder time whenever prices dropped. Nationwide, forty acres of ponds appeared to be enough to provide a livelihood for a fish farmer in 1968, but that minimum increased to one hundred by 1970 and appeared to be as high as three to four hundred acres by 1985. Farmers with larger farms could benefit from economies of scale.[1]

Some observers attribute farmers' staying power to their financial resources. "The only people who've stayed in," says one farmer, "were the millionaires who could afford to survive the crashes. They know that after the crash the prices are gonna go back up again, and if they can hold on long enough and they've got pretty deep pockets, they can take advantage of the next up-cycle. So they stay in."[2]

Farmers able to make good business decisions had a clear advantage. For some, the decision to start producing was not well informed, and they did so without having a certain market. Many succumbed because they could not sell their fish. Many speculators entered the business without understanding the risks involved in catfish farming. While some farmers were able to increase the level of mechanization and scale of their operations in order to stay ahead, others were unable to make ends meet. Some, such as Dan Butterfield and David Pearce, were able to weather times of low channel catfish prices because they grew other fish species in addition to catfish.

Sometimes, it was a matter of luck. Some markets on which growers depended fell through or did not live up to expectations. Consumer demand was limited for years by seasonal availability, by off-flavor, and simply by the fact that many consumers still thought they would not like catfish. It took a long time and many people working together for demand to increase. Even in areas where market demand was great, many farmers had borrowed heavily and were burdened by high-interest loans. In many cases, those farmers went out of business because they were unable to make payments and, as a result, lost their land.

Economists generally agreed that how farms were managed was becoming the single most important key to fish farmers' economic survival. "Skillful, efficient management," both from an aquacultural as well as an economic standpoint, was considered essential for a farm to be profitable. Successful farmers learned from their experience farming and running the business:

"The people who have been successful know there are few textbook answers, that every inch of earth and every drop of water is different. Even ponds, side by side, react differently and fish production will vary in the same pond from year to year. Those who had made money were those who had almost nothing with which to begin. Their mistakes were small ones from which they learned."[3]

When the economy began turning upward in the late 1970s and the 1980s, there were fewer catfish operations. Many were larger than before. The growers who remained raised more fish in more ponds and made more money from sales. The comeback lifted Alabama from third position to second below Mississippi in production. Farmers had been figuring out ways to operate more efficiently; paddle wheel aeration had boosted the carrying capacity of ponds; options for transporting the fish were improving; and processors were putting more energy into finding new buyers for their products. Promotion efforts were paying off. The demand for catfish products was growing all over the country, and exports abroad were opening up additional markets. As a result, processors began to pay farmers better prices for their fish; farmers began producing more; and the industry started to show vitality again.

Although the improving trend prompted yet another stampede into fish farming around 1980 and created another oversupply, Joe Glover's voice served as a reassuring influence on Alabama farmers. He encouraged them to be patient and to accept the ups and downs that take place in the pricing of all commodities. He communicated to them his own confidence that the industry would even out again. His optimism helped many farmers through.

Gradually, a manageable balance between demand and supply evolved, and price fluctuations became less dramatic. The industry began to settle down. Farmers were finding ways to deal with farm employment difficulties. The federal government gave recognition to fish farmers by passing the National Aquaculture Act of 1980 and adopting the National Aquaculture Plan, both of which prompted and supported important agency assistance that followed. A cooperative outreach effort was established and institutionalized in the form of the Alabama Fish Farming Center. Producers banded together into associations to gain some control over economic conditions and accomplished many of their goals.

People had hoped that catfish farming would create much-needed jobs for the area's residents. Employment opportunities on Alabama's early catfish farms were limited, however. Farmers generally fed and oversaw the fish themselves, often with help from family members. Additional hands were usually needed only seasonally or temporarily for hatchery work in May and June, for moving fingerlings between ponds, for nightly oxygen checks in late summer, or in fall or winter if a farmer did his own harvesting. As processor harvesting and year-round stocking became the norm, the need for seasonal help on farms decreased.

Finding occasional or temporary part-time help just when they needed it was not always easy. Hiring through acquaintances provided some assurance of employees' reliability, but catfish work was not attractive to many people, and many did not stay long with a farm. Even today, employees may want to work longer but not want to commit themselves permanently, according to Bill Kyser. They might work for six months and then move on.

Many farmers found ways to adjust their farm activities to the availability of extra hands. Some farmers could keep workers year-round by employing them in their other farming operations when they were not needed for catfish. That way, someone would always be available when help was needed. Planting and harvesting row crops or baling hay, for instance, were done at times of the year when ponds required less attention. Harry Miller and his younger brother, Neal, went into the bee business partly because shipping the bees and honey could be done in the winter, providing work for their farm labor the year round.

Problems of training and skill levels were associated with temporary labor. Even if they worked only part-time, employees with the knowledge and training to do their jobs became highly valued. This is still true today. "Part-time help is untrained help, and it places certain limitations on you," explains hatchery owner Bill Easterling. "An inexperienced person creates problems for the workers who are experienced. I always tell a person, 'If you come to work, it takes six months for you to really be an effective worker for us.' We don't have sufficient funds to train a person. We have to train them on the job, and there's a lot of correction—correcting things that they do wrong." Considerate management can result in less transiency, according to Easterling. "We try to prepare unskilled employees mentally so as not to embarrass them or hurt their feelings in the process of correcting them. It has worked real good. Most of the ones that have come to work for us work several years."

Paying salaries and fringe benefits comparable to those offered by many corporations has helped some farmers attract and keep valuable employees. "We organized things trying to make it much more livable," says David Pearce, "some fairly good benefits, a profit sharing plan, insurance, with the whole idea of trying to hold onto the people that we have." In the early days, when most of the heavy, physical labor was done manually, according to Bill Kyser, "we tried to keep one man around, but it was hard to find somebody and keep them. In the last ten, fifteen years, we have been able to find some pretty good help that we've been able to train and keep interested. We have to pay them considerably more than minimum wage, but they need more than that." But higher pay and benefits were something usually only farmers with larger operations could offer.

The industry had undergone two decades of development before outreach to fish farmers and would-be farmers became a concerted effort in Alabama. Because of its focus on international efforts, the AU fisheries faculty had at first not been very involved in domestic catfish farming. Extension involvement in the industry leaped forward in 1979, when John Jensen, who had worked in the Peace Corps in Brazil, became one of the first Alabama Cooperative Extension specialists in aquaculture. Jensen worked energetically to share technical and economic findings with farmers, and before long, he was spending close to one-third of his time just answering their telephone calls and letters.

Not only farmers benefited. Faculty members who had taken time from their research to answer farmers' requests were able to focus on research again. Jensen's consultations with professors stimulated their interest in what was going on in the state's catfish industry. They focused more of their research activities on the problems Alabama farmers were facing.

The National Aquaculture Act of 1980 made the federal and state Cooperative Extension systems expressly responsible for outreach to fish farmers. But county extension agents had little or no training in aquaculture, so Jensen undertook to prepare agents for their new responsibilities. Going from county to county to present hands-on aquaculture workshops and meetings, he trained virtually every agent in the state. They were then able to take fisheries information to the people in the counties, to talk the catfish business with them in coffee shops and in grocery stores, and to give them assistance at the ponds.

Help came also from other sources at Auburn University. Research economists such as E. W. McCoy began to examine economic aspects of the industry, building on limited work done elsewhere. They began to look at fish farm planning, management, and marketing. Starting a commercially viable catfish farm was a major capital investment already in the 1970s, and individual farmers needed to be able to make informed decisions regarding whether to enter catfish farming and ways to keep their farms profitable. Extension economists such as Jerry Crews responded to the needs primarily of small-scale producers by adapting enterprise budgets used previously in connection with other farm commodities to their situations.

During the 1960s and 1970s most of Alabama's fish farmers were either unaware or did not understand much of the information that was available about their problems. Farmers who sold fish to Farm Fresh Catfish Company could consult with company biologist O. L. Green. Others turned to the SCS. Neither were able to provide all the help farmers needed, however. Many turned to other growers for guidance, resulting, at times, in a situation of "the blind leading the blind."[4] Few other resources were available besides the Cooperative Parasite and Disease Project, which was hours away at Auburn.

Ernest "Ernie" Todd, head of the state SCS, knew something had to be done. Recognizing that establishing a cooperative organization to help the farmers had to have broad-based support in order to succeed, he coordinated the efforts of many individuals and groups working to establish a large-scale, cooperative resource for the farmers. Together, he, Dave Kelly, Elvin Wright, local officials in Greensboro and Hale County, and officials from the SCS, the Alabama Association of Soil and Water Conservation Districts, the Alabama Cooperative Extension Service, and the Agricultural Experiment Station at Auburn, eventually got approval and money. In 1982, the legislature granted two hundred thousand to establish the Alabama Fish Farming Center.

The county loaned an old farmers' market on Highway 69 in Greensboro rent-free to the local Soil and Water Conservation District. Inmates from the nearby State Cattle Ranch correctional facility did most of the construction. The building was transformed into office and laboratory space, with additional labor and materials furnished by the City of Greensboro. The center was staffed with technical experts and support personnel—an SCS engineer to help design ponds, an extension aquaculturist from Auburn to help farmers manage ponds, and an experiment station fish health specialist to diagnose and treat diseases. By 1983 it was ready to serve West Alabama's fish farmers.

The farmers were gratified and encouraged. They no longer had to send diseased fish to Auburn. They could bring fish and samples of pond water for analysis to the center's lab, and, when needed, center staff conducted analyses at their ponds. In its first year, the center served two hundred farmers and eight thousand acres of ponds. The staff held workshops in which farmers and prospective farmers could learn basics of farm and hatchery operations. Farmers took full advantage of these opportunities, attending workshops sometimes in numbers well over a hundred. The center also organized conferences and trade shows, providing a forum for ideas and discussion about all phases of the industry, not just farming. Perhaps one of the most significant educational roles it undertook was teaching farmers about the relationship between intensive production and water quality degradation. Helping them strike that balance in their ponds was one of the most difficult yet crucial tasks center aquaculturists had to undertake.

As the industry became established, growers felt the need to exert economic and political influence to protect their interests. Mississippi farmers had organized themselves in 1968 to form the Catfish Farmers of America (CFA), the largest trade association for freshwater fish. David Pearce, Harry and Neal Miller, and other West Alabama farmers agreed in 1972 to form an association, the Alabama Catfish Producers. "We realized we needed to have a forum," says Pearce, "for farmers to come together, talk about the problems they have, and try to present a united front to the research organizations." Official support for the industry was not yet a state priority, however, and the organization operated on its own for two years.

The Alabama Farm Bureau Federation had been representing Alabama farmers' interests since 1921 and had begun to reorganize along the lines of various agricultural commodities such as beef, pork, and cotton. Alabama farmers benefited from the bureau's support and advocacy. When Pearce and other catfish farmers approached farm bureau officials in 1974, the officials agreed to take the group under the larger organization's wing and added a catfish division. A ten-member State Catfish Committee, composed of farmers elected by other farmers for a year at a time, was given a small budget to cover expenses and, with Pearce as its first head, began to identify actions that might best solve the farmers' collective problems.

The farm bureau's resources had been focused mainly on the needs of producers of traditional crops. By 1978, however, the catfish industry had developed solid roots in Alabama's economy, and after a membership fee increase, Jimmy Carlisle was hired as director of both the poultry and catfish divisions. He began to devote his energies increasingly to the needs of the younger industry, which seemed more pressing than those of the better established and more mature poultry industry. Although the name and ties of the organization changed during the late 1980s, its support of the farmers never wavered.

Carlisle and John Jensen undertook many projects to help the state's farmers. In the 1980s they composed an industry newsletter, and Carlisle started an industry publication, the *Alabama Catfish Producer.* They organized statewide trade shows and the first Alabama Fish Farming Conference. The conference, which allowed farmers to exchange information and to benefit from research and technological advances in the field, was so successful that more were held in following years. After about ten years, however, the state catfish committee decided to sponsor trade shows jointly with the three other top catfish-producing states—Mississippi, Arkansas, and Louisiana—in order to avoid duplication and save money, time, and effort.

In the late 1980s the catfish division became the Alabama Catfish Producers' Association, in line with similar associations in other states. Its structure and responsibilities stayed basically the same. "We will continue to examine current regulations and policies that have certain constraints on aquaculture and then work with the various agencies to make desirable changes," announced the *Alabama Catfish Producer* in April 1987.

The farm bureau's legislative branch had people who knew how to write bills. They wrote and lobbied, together with the catfish committee, for legislation that significantly improved conditions for farmers. The laws usually moved swiftly through the legislature with the support of representatives from local districts in which the industry was a major economic player.

One legislative action of the committee and the bureau in the 1970s involved the "prompt pay" law. Early on, farmers sometimes went for long periods of time without being reimbursed by processors, who were reluctant to pay immediately for fish they had harvested unless they felt certain they would be able to sell them. Similar situations in traditional livestock sales were already covered by the Alabama Livestock Dealers' Financial Responsibility Act of 1969. Farmers selling live cattle, horses, and other farm animals had to be paid the next business day after the sale. No such law applied to catfish. To help producers get fair play from processors, the bureau lobbied to have the definition of livestock in the existing act changed to include catfish. In 1984 an amended version of the "prompt pay" legislation passed, providing guidelines for paying farmers within ten working days of harvesting.

~

> The Brazilian catfish—if indeed it actually can be called a catfish—is being sold throughout our country as a farm-grown fish. When you bite into it, you will wish you had never heard of pond-grown catfish. This Brazilian fish is doing more damage to the image of the pond-

> raised catfish than can be corrected in ten years of promotion by catfish farmers.
>
> —Roy Prewitt

Another early legislative project involved the Alabama Catfish Marketing Labeling Act. Imported Brazilian river catfish had become an increasingly aggravating problem. Although they belonged to an entirely different family of catfish, they were marketed without that distinction being made. Alabama's farmers and processors, many of whom had small operations, found themselves competing with imports that sold at a much lower price than what they could afford. Concerned about economic survival, farmers asked the state catfish committee for legislation.

An existing labeling law required identification of crude protein content on feed labels. Carlisle, together with the committee and bureau staff, drafted a bill that required identification of farm-raised catfish, as distinct from imported river fish, on packaged products. If consumers had that information, they reasoned, they would choose to purchase the farm-raised, domestic product. The Consumer Marketing Labeling Act was passed in 1975.

> People are stealing all of his fish from him, and there's nothing he can do but watch them. It's just not fair. When a farmer starts to genetically improve his stock, he has money invested in that improved fish. Most people would not make a distinction between a superior broodstock and your normal everyday recreational fish. When people came in and stole those fish, they were stealing a livelihood from that farmer.
>
> —Jimmy Carlisle

Poaching had been a major problem for fish farmers since the early years of the industry. It was a concern nationwide and abroad. Fish farming journals of the early 1970s devoted whole articles to the problem. Farmers simply could not keep a twenty-four-hour watch over all their ponds. J. V. Friday remembers his poaching troubles starting with his fee-fishing ponds: "We started with one pond. Then I began to let people go a little further out . . . a little further out . . . and that's when I began to lose my fish. They'd slip in and take them or drive out without paying for them at other ponds where they'd be out of sight. I'd just as soon quit the business."

By the early 1980s "catfish rustling" was a significant problem for some growers. This was not a matter of individuals engaging in subsistence fishing with rod and reel. It was a multimillion dollar-a-year, illegal enterprise in fish-farming areas. One West Alabama farmer lost twenty thousand pounds of broodstock catfish in three months that were sold to a restaurant supplier. The broodstock were between four and eight years old, and at the prevailing value of ten dollars per pound, the thefts amounted to a two-hundred-thousand-dollar loss.

The solution, as farmers saw it, was to get strong state laws making catfish poaching a crime. Where applicable laws were on the books, local officials typically failed to enforce them. Alabama's catfish farmers wanted stricter enforcement with stiffer penalties. In August 1982 state legislative representatives from catfish farming areas joined with others to lobby for a new antipoaching bill. A bill was passed the next spring establishing catfish theft from ponds as a criminal offense, with stiff fines for anyone convicted. First-time offenders could be imprisoned at the court's discretion, but jail time on later offenses was mandatory and fines harsher.

The checkoff program was perhaps the most far-reaching action taken by the catfish committee. In an early effort to organize a self-taxing system in the 1970s, West Alabama farmers had attempted to collect fifty cents on every acre of ponds to deposit into a fund for their organization's activities. The fund proved inadequate, however, and the effort stalled. The idea of self-assessment floated around among producers and state committee members for years and finally gained grassroots support again in the mid-1980s.

This time, the idea was that farmers would pay a certain amount of money per ton of fish feed they purchased. With this income, they would be able to pay for projects benefiting and promoting the industry. In Alabama, any sales tax required a statewide vote to pass, even if members of the group paying the tax had already agreed to assess themselves. Jimmy Carlisle and the committee had to place the issue before the voters in a public election. In 1987 the checkoff went before the state's voters for their approval, and the program passed the referendum.

The Alabama Farm Bureau had recently withdrawn from the American Farm Bureau Federation and become the Alabama Farmers Federation. The catfish division, like the other commodity divisions in the organization, had very limited funds. The checkoff program provided the financial means to help the industry develop. In July 1989 catfish farmers began paying two dollars per ton of feed over and above state sales tax. Over the years, checkoff monies went to support, among other things, research, outreach, a quality assurance program, a state aquaculture plan, catfish festivals, aquacultural education in the schools, and media promotions. Word of the program spread to neighboring catfish-producing states. Industry representatives from Louisiana and Arkansas contacted Carlisle. Both states eventually modeled similar actions of their own on the successful Alabama program.

In 1993 the checkoff program came up for renewal. The program had been extremely successful, collecting enough money from farmers for the organization to give out hundreds of thousands of dollars to various projects over the years. The producers' association drew up a new referendum that went in 1994 to the voters, who passed a new assessment of only fifty cents per ton. The assessment was reduced partly because earlier that year, the state association and West Alabama farmers had agreed to support generic advertising done by the Catfish Institute in Mississippi instead of conducting Alabama-specific promotional campaigns. Auburn University economists had determined that generic advertising was a more efficient use of their money.

Bill Easterling's fish hatchery was the oldest, largest private hatchery in East Alabama for catfish and other species. For about ten or fifteen years some of his ponds were open to the public for fee-fishing. People would come from all around, including Georgia, lugging picnic gear, chairs, and fishing poles of all types and sizes to catch as much out of his millpond as they could in one day. Or for the same fee, people could choose to fish for one hour in the "bonus raceway," where fifteen hundred gallons of water per minute flowed through the pond and supported many more fish. On Saturdays and holidays and in summertime, large numbers of people came regularly to fish from its banks.

One day in 1986 or '87 state game wardens walked up to the pond. There were two or three of them, according to Easterling, and "they went around our raceway, checking fishing licenses. Several people from out-of-state were fishing, and the wardens gave them a citation to appear in court." Easterling was in a disturbing predicament: "We had invited people, charged them to fish, and then they were fixing to have to go to court. I told them that we would defend their case in court. I had taken pictures of the game wardens as they were giving the tickets out, which they were not too pleased with. The judge was very considerate. He told them if they would buy a license, they would not have to pay a fine."

This situation highlighted clearly to Easterling the need for a change in the law. Other farmers in the state were encountering similar problems at their fee-fishing ponds. They talked the situation over with members of the state catfish committee. "If a farmer grows fish and stocks them in one of his ponds, why does the public have to have a state fishing license when the state of Alabama never had ownership of those fish in the first place?" they asked. They felt that fee-fishing should be viewed as "marketing a farmer's product like marketing strawberries and marketing tomatoes," explains Jimmy Carlisle. The fish belonged to the farmer who had raised them, and he ought to be able to sell them to anyone in any way he chose. They went to the head of the Division of Game and Fish and to other state officials, all of whom agreed. The law regarding state fishing licenses simply did not fit the privately owned, commercial fee-fishing situation.

The state committee proposed an exemption to the law: if growers would post a sign identifying their ponds as farm-raised catfish operations, customers from any state could come and fish at any time and would not need a license. After working cooperatively with state Fish and Game officials on drafting it, the committee saw the legislation pass relatively easily in 1987.

The double-crested cormorant is a waterbird that is highly adapted to both freshwater and ocean environments. It is a "generalist," foraging in virtually any open water. Its webbed feet make it an adept diver that eats mostly small fish it chases underwater. Bare orange patches on its face and throat stand out against its blackish body. The most common cormorant in North America, the double-crested's populations were dramatically reduced by DDT and other toxins during the middle of the twentieth century.[5]

The great blue heron is a large wading bird, the most widespread and familiar of the herons, with a long neck and long, stiltlike legs. It is very large, with gray plumage and a black stripe down the top of its whitish head. Its massive, pointed bill is adapted to spearing and grasping prey, which it catches as it stands patiently at the water's edge. The heron's prey may be fish or other aquatic or semiaquatic creatures. Visitors to natural aquatic ecosystems around North America enjoy seeing these beautiful and graceful creatures.

In contrast, the sight of the cormorants and herons sparks dread and frustration in the hearts of commercial catfish farmers. Once migratory or only occasional residents in West Alabama, the birds have established burgeoning new populations there, attracted by a continuous and plentiful source of food—young catfish. "Before we grew catfish," says Joe Glover Sr., "it was a very rare thing to see great blue herons and cormorants. Cormorants can swim under the water faster than the fish, and they are a protected species. Those birds have always been considered migratory. Now they are nesting and living the year-round. We have them by the hundreds of thousands." The situation is expected to get worse. The herons have shown a remarkable ability to adapt to their new feeding grounds with new behaviors. "Great blue herons are not going to leave," says David Pearce. "You can run them off, and as soon as you leave, they're right back. Each one eats a minimum of two-thirds of a pound of fish a day. We got that documented. They're a wading bird. We've got them on video tape landing right out in the middle [of the pond] where we're feeding fish, swimming like a duck and just picking the fish up. It's not hard to count them. We've had as many as a thousand on the place at a time."

In recent years, cormorant populations have increased exponentially. The once-migratory birds are protected under the Migratory Bird Treaty Act, and farmers had limited options with which to deal with them. They petitioned for help through their national organization, the Catfish Farmers of America. Relief came from the federal government through Wildlife Services in the USDA's Animal and Plant Health Inspection Service in 1997 and 1998. Legislation now allows farmers to use scare devices and to shoot cormorants or other birds if they are seen in the act of taking fish from a commercial catfish pond. In addition to cormorants and herons, growing numbers of egrets, pelicans, and even wood storks have begun to feed at the ponds year-round or while overwintering.

Catfish farmers have a variety of problems with wildlife on their ponds. There have been incidents of muskrats burrowing into levies and weakening dams. River otters have gotten into ponds and eaten fish, and on occasion, harvesters net alligators in their seines. For these and for bird depredation problems, farmers can get assistance from the USDA Wildlife Services Agency. Wildlife Services director Frank Boyd works closely with Alabama Division of Wildlife and Freshwater Fisheries biologists and with university researchers to help the farmers. He helps assess problems and helps design and set up control programs as well as research and test control methods such as noisemakers and other devices. The Wildlife Services Agency now recommends "an integrated scaring program with a wide variety of scare devices supplemented with limited lethal control," according to Boyd. Just as with disease and water quality, farmers have learned that no single action will solve their predation problems once and for all. Rather, predatory bird control is becoming accepted as an ongoing part of catfish farm management.[6]

Commercial fish farming developed unevenly around Alabama. By far the greatest concentration of production was always in the west, but commercial operations exist in almost every part of the state. Farmers in the various regions formed their own associations in order to exercise control through strength in numbers over their different circumstances.

One of the primary issues on which regional organizations made significant inroads was that of feed prices. Alabama has many feed mills now, and competition among them keeps the price of fish feed down. But that was not always so, especially during the early 1980s. Farmers in West Alabama could not offer as large a market for feeds as could the much larger farms in Mississippi. As a result, they were paying far more for feeds than were growers a few miles away on the other side of the state line.

The West Alabama Catfish Producers Association was organized around this issue in 1984. At one of the group's first meetings, the farmers agreed to act together to bring down the price of feed. They combined their separate feed orders into one large order and submitted it to the four or five Alabama mills for bids. None of the feed manufacturers wanted to lose the single order for fourteen thousand tons and be left out of the largest catfish feed market in the state. In order to break up the farmers' buying cartel, the mills announced that they were lowering their prices significantly to a level roughly equaling prices in Mississippi.

Gratified by their victory, the farmers then passed word among themselves for each of them, independently, to buy the feed he needed from any of the mills. "Just go ahead and buy it. Everybody buy," was the message that went out. Their action helped save all the manufacturers and at the same time got prices down to a satisfactory, competitive level. "We felt that by joining together, West Alabama catfish farmers could best deal with the problems facing their industry," commented Thed Spree in 1985, president of the organization at the time. "Since the industry is centered in West Alabama, we've got the strength that's needed to get things done."[7]

The group's successful beginning propelled their association on for years of organized activities such as educational meetings, at which Auburn University researchers gave talks, and social and promotional events such as picnics and fish fries. It even launched a successful 1989 lawsuit on behalf of a member farmer. Eventually, as the supportive role of the Farmers Federation grew, the West Alabama association became less active as a distinct, regional entity.

Farmers in other parts of the state wanted to strengthen the industry in their areas. In East and North Alabama, they formed their own organizations in the hope that it would help them get processing facilities in their areas and find ways to sell their fish. Little progress was made, however, and the organizations became inactive. The Piedmont Association of Caged Fish Producers still functions in the northeastern part of the state and directs its energy to supporting and promoting the cage industry.

Over the years, producers' needs for the Alabama Fish Farming Center's services grew faster than the staff's ability to cope with them. In the ten years after its creation in 1983, the center's staff grew from three people to seven. The pond acreage served grew from eight thousand to eighteen thousand, and the number of farmers served nearly doubled to five hundred. The number of disease and water quality cases the staff handled tripled. The center's advice was saving 10 million pounds of fish a year.

The existing physical facilities and operating funds were inadequate. Those who recognized the importance of its mission had joined with local legislators earlier to lobby for increased support but had little success. In the fall of 1990 retired Hale County probate judge Richard Avery decided to change that. He undertook a campaign to fund and build a new center. Convinced it could be done with grants, donations, and in-kind services, he approached local and state agencies for permission and support. Eventually, the Hale County Industrial Development Board agreed to sponsor the project, which then received a state grant. Inmates from the State Cattle Ranch worked once again for the catfish industry as carpenters, masons, and cabinetmakers. The massive cooperative effort among community businesses, organizations, individuals, and local, county, and state agencies resulted in enough contributions of funding, labor, and materials to complete construction of a new, larger facility in 1992.

Before long, the center was diagnosing and assisting with close to eight hundred disease cases and 1,750 water quality cases annually. Its engineer was designing, on average, 560 acres of new ponds a year. The center was unable to pay for its growing caseload. Its umbrella agencies were unable to provide more money, because their own funding had not increased nearly as fast as the industry had grown. Checkoff funds came from the state catfish committee only periodically. In 1996 the Alabama State Soil and Water Conservation Committee found a way to significantly increase the amount of money it contributed to the center. Within a few years, the center finally began to receive enough funds to do the work the farmers needed. The Alabama Fish Farming Center's establishment was a milestone in the development of the state's industry, and the continuing, adequate support it began to receive was pivotal to the industry's advancement.

Notes

1. Just as grocery store items in larger packages usually cost slightly less per pound than the same items in smaller packages, farmers raising more fish benefit from economies of scale because they buy larger quantities sold at cheaper per unit prices.

2. Dan Butterfield, interview by author, tape recording, 14 January 1998, Special Collections and Archives, Auburn University, Auburn, Ala.

3. Prewitt, "Rambling Along" (1970): 24.

4. Dave Kelly, "The Alabama Fish Farming Center," *Alabama Agribusiness* 22 (July 1983): 4.

5. David Allen Sibley, *The Sibley Guide to Bird Life and Behavior.* (New York: Alfred A. Knopf, 2001), 160–62, 164.

6. Frank Boyd, telephone conversation with author, 10 December 2002.

7. Quoted in Jay Burnett, "No Way to Go but Up." *Alabama Farm Bureau News* 61 (spring 1985): 6–8.

15
More Processors

> The first thing I had to make sure that we did in the beginning was survive. My goal was to reach a hundred thousand pounds of live fish a week and maintain that. I never dreamed that this business would grow to where it is today.
>
> —Joe Glover Jr.

The majority of the processing ventures started through the years are no longer operating. Small operations were typically established by farmers after they found that local residents and businesses were eager to buy processed fish. Many could not produce enough even to satisfy the demand in their areas.

One of the smaller businesses still operating today began around 1966 on a picnic table by the side of Lester and Milton Taylor's fishout pond in Town Creek. Cleaning catfish caught by their customers, the Taylors found sales went so well that in 1970 they built a small fish market next to some of their ponds, with a processing facility in the back (see fig. 18). Like the Taylors, other pond owners in the area were unable to lure processors' harvest crews from Greensboro to northwest Alabama, so they gladly agreed to let the Taylors harvest their ponds. Before long, the Taylors were buying fish from as many as twenty other producers. They began a contractual relationship with the pond owners. They stocked the ponds, furnished feed, and harvested the fish. In return for getting paid per pound harvested, the farmers fed the fish and maintained the ponds. The arrangement worked well. Residents and visitors to Town Creek can still buy fish at Taylor's Fish Market today (see fig. 19).

Like the Taylors' operation, the Fridays' processing business in Selma developed out of their fee-fishing enterprise. J. V. Friday started raising catfish in 1966, and his fish-farming business was a local hit from the very start. "Back then everybody wanted to fish," Friday says, "and it was a cheap place for them to get meat. It was nothing to have a hundred people there at a time. Fishout was mostly on Friday and Saturday and, in the summer,

Fig. 18. Unloading harvested fish into holding tank, 1998.
Harvested fish are unloaded from a truck into holding tanks for hand-processing in the back of Taylor's Fish Market in Town Creek. (Photo by author.)

Wednesday." But it was time-consuming for Friday. "You had to stay there. When I was staying there with customers, in place of just sitting there holding my hands, I got started dressing fish, just to have something to do. I had a refrigerator, and I'd put them in there in individual bags. One thing led to another. . . . "

Dressing the fish in addition to operating their beef cattle farm, catfish farm, and fishout ponds, the Fridays had more than they could handle. Many of their pond fish were being lost to poachers, who came on off-days, when Friday had to be away. So in the early 1980s the Fridays closed their fee-fishing operation and converted a small cement block building next to the ponds into a processing operation. Friday, his wife, Ellen, and their son did the work themselves. Friday went door-to-door to local businesses—restaurants, grocery stores, the hospital—and placed advertisements in the newspaper. Within ten to fifteen years, it seemed as though they could not raise and process enough fish for the many customers who wanted them. Local residents stocked up on catfish two days a week, and the rest of the

Fig. 19. Lester Taylor, early 1970s.
Lester Taylor stands in front of Taylor's Feed & Fish Market. The feed company that donated the sign added the words "Feed &" to the sign. (Photo courtesy of Nellie Taylor Estate, Milton Taylor, Personal Representative.)

time Friday and his family were busy raising, processing, and delivering twice weekly to local markets, restaurants, and hospitals (see fig. 20).

Dan Butterfield established his farm near Tuscaloosa in 1969. On days when he opened for retail sales, he harvested fish and stocked them in holding tanks for customers. Butterfield saw, after his first year or two in farming, that very few customers wanted to clean the fish themselves. In order to market his fish, he realized he would also have to dress them. So he

Fig. 20. The Friday family's catfish trailer, 1998.
J. V. and Ellen Friday assembled the trailer themselves. With it, they delivered fresh catfish twice a week to local businesses and institutional customers such as restaurants, hospitals, and small grocery stores. The trailer held approximately two thousand pounds of fish. (Photo by author.)

learned how and was eventually able to claim a record hand-dressing time of thirteen and a half seconds. With few overhead expenses from his outdoor, pondside market, he had great success selling the fish (see fig. 21). Because there was very little competition in the area, people came from as far as a hundred miles in all directions for the catfish and other species he grew.

The bulk of processing that took place was in larger plants such as Farm Fresh. In spite of all its early difficulties, Farm Fresh grew steadily. Optimism and determination helped Joe Glover Sr. work his way around the many obstacles he encountered. He worked tirelessly promoting the plant's products, and sales increased. Glover had built additional ponds on family property and had obtained financing to buy more land so he would be assured of fish when farmer's supplies fell short. Yet Farm Fresh continued having difficulty getting enough fish throughout the year. Because Alabama farmers were not yet raising enough, he began to look across the state line, where large farms were producing many more fish. Few processing plants in

Fig. 21. An A-frame table for hand-dressing fish, 1998.
The table stands next to ponds at Dan Butterfield's fish farm in Tuscaloosa. (Photo by author.)

Mississippi had been successful at marketing their products, and Glover was able to purchase one in Hollandale in 1972 and another later in Arkansas.

In the meantime, ConAgra had been operating the old STRAL plant right across the street from Farm Fresh, but its stay in Greensboro was temporary. In 1975 the company pulled out of Alabama to focus its resources instead on its two Mississippi plants. The Greensboro facility went back to local ownership when Richard True, Joe Kyser Jr., and Clint Wilkinson, a Selma dentist, bought it the following year and began processing under the name Country Fresh. True was back managing the plant he had originally co-owned and was now competing with his former partner across the street. Kyser, who had gotten his start at ConAgra's hatchery and on its harvest crew, oversaw procuring and harvesting.

Country Fresh's operation progressed slowly at first, but the business did well. When "we bought the plant, we had two old trucks," remembers True. "In two years time, we had thirteen trucks, and they were all paid for." The original STRAL plant had been built large enough to process eight to ten thousand pounds a day, but the facility was no longer adequate. In 1980 when he retired from Country Fresh, True says, "we were processing thirty-five thousand pounds a day and employing forty or forty-five people." Country Fresh continued to operate for about two more years, at which time Joe Glover Sr. bought it and closed it down, leaving Farm Fresh the only large processor in Alabama.

In 1983 Glover sold Farm Fresh's processing business to George A. Hormel and Company, one of the large corporations attracted to the promising catfish enterprise. He stayed on as president of the company for five years but then retired. Hormel had leased and operated all his Farm Fresh ponds, but when the lease ran out after twelve years, Glover became a catfish farmer again.

Hormel, a nationwide company from outside Alabama, kept the Farm Fresh name, but the switch in ownership brought changes in procedures that did not always go over smoothly with the processing line workers. Alabama winters can be fairly cold, but when Hormel took over, the heaters that had been in the plant were removed. Heaters were not allowed in Hormel's meat processing plants. The edict to remove the heaters almost caused an uprising among the employees. Randy McCrory, who was by then a manager, handled it as best he could. "We told them, 'We're sorry, but those are the rules and regulations initiated by Hormel. [You've] got to abide by them, or else find yourself another job.' We tried to handle it as easy as we could and explain why." The only solution was to dress more warmly, he told employees. "You're going to have to wear more clothes, just like I do myself. Come prepared to work in a cold environment." McCrory and the workers wore sweat suits and thermals under their smocks and then shed layers as they warmed up working. It was difficult at first, but most eventually adjusted. Those who did not, quit.

The new owners introduced certain other procedures used in Hormel's meat-processing facilities. These procedures became standard in catfish processing. Workers were required to wear latex gloves, which were more sanitary than the cloth ones they had been using. The combination of the cloth ones under the latex ones kept their hands warmer in the wintertime but made it more difficult for employees to manipulate objects. Over time, improvements in thicker protective gloves made for easier handling of the fish. McCrory recalls that Farm Fresh "always tried to find what was best for people as well as for the job."

Joe Glover Jr. (Joe Jr.) had worked at Farm Fresh while still in high school and spent as much time there as he could. His father had learned everything there was to know about the business by trial and error and was determined his son should learn it all as well. He knew that in order to get respect and be able to tell others how to do things, one needed to know how to do the job oneself. So Joe Jr. worked in all phases of the plant's operations. He learned about harvesting, processing, and marketing. He cleaned fish and cleaned the plant. His father made certain he did everything so he would understand each particular job, the flow of the product, and how to run a business.

Joe Jr. continued working at Farm Fresh after it was purchased by Hormel. He had high hopes of accomplishing much in the industry in Alabama, but Hormel had its own agenda. After one year, his position was eliminated, and he received a cut in pay. It was clear to him that the company's interest lay more with its Mississippi operations than with the plant in Greensboro. Both father and son were acutely aware that Farm Fresh had made a commitment to the area's farmers to do what it could to help them market their fish, and both felt the obligation of that commitment.

The time seemed right for Joe Jr. to strike out on his own. In July 1985 he left Hormel and started construction on his own plant. He had saved some money and was in a little better financial position than his father had been when he had started Farm Fresh. He had $115,000 in savings and was able to get a low-interest agricultural development loan for $250,000. Within half a year, Southern Pride Catfish Company was ready to process fish.

Just as his father had in the early days of Farm Fresh, the younger Glover had to struggle to get his business on its feet. The plant was not large, but no money was left over to start operating. He had raised some fish on his own farm and used some of them now, because he could delay paying himself for his fish longer than he could put off paying farmers for theirs. This gave him some time and allowed him to generate some money with which to begin buying fish and paying employees. After two or three months in business, he found himself again without money to pay his employees. Southern Pride still needed to get customers and to generate cash flow. With the sale of the home in which he and his wife had previously lived, he bought more time.

Before long, however, Joe Jr. was once again unable to make payroll or to pay for live fish. He was fast gaining experience in keeping a positive attitude. "When you're going through this, you still have to be positive and motivate your people and not let them know you're having financial problems," he recalls. "We survived, but those first four or five months were really scary. When I left the plant at night after sixteen hours, I wondered many nights whether I'd made the right decision to go in[to] business for myself. . . . I did not have adequate funds to be doing what I had set out to do. I knew I'd have to have help from somewhere. I did not want to sell stock to generate money because I wanted it to be a family business. But I didn't know what I'd have to do to generate cash to operate. That's the scary part, not knowing where the next dollar is coming from."

Having worked for his father, Glover had seen times when there had been cash flow problems, and he knew what to expect. He knew, too, that things could get better, so he kept working and hoping. Gradually, he began to see signs of a profit. After the first year, the plant had established a solid base of customers, and there was money to operate the business. Glover was able to continue fulfilling his father's commitment to sell the farmers' fish.

Glover had gained a wealth of useful experience at his father's plant. One of the keys to operating a successful business, he had learned, was to surround oneself with people who were self-motivated, who showed drive and ambition. Southern Pride had begun its first day of processing on January 13, 1986, with 35 employees. The plant ended its first year having processed 7 million pounds of fish and employing roughly 150 people.

Until the establishment of Southern Pride, Farm Fresh had been operating as the only large processing plant in Alabama. With only one plant buying fish, many farmers were reluctant to make sizable investments in fish farming. The existence of a second plant gave them a choice of where to sell their fish and boosted their confidence in the industry's future and, therefore, in their own. Production increased. After Hormel closed the Greensboro plant, all significant competition for Southern Pride disappeared until the establishment of another plant in 1991.

In 1994 Southern Pride expanded to Demopolis, where it had purchased a plant originally built by Blue Waters Processing Company. The Demopolis plant soon employed roughly one-third of the company's total workforce and processed an equivalent proportion of fish daily. Southern Pride is now the world's largest processing plant in the catfish industry.

George Smelley worked for the SCS for ten years after graduating from Auburn University in 1981. Because a major part of his job was designing ponds in Hale County, he was able to watch the catfish industry develop. "I learned it from one end to the other," he says. Smelley felt that catfish farming was a perfect enterprise for him.

Many people were having a difficult time farming in the mid-1980s, and land prices were low. It seemed to Smelley to be a good time to start farming on his own. With the money from the sale of his house, he made a down payment on some land that he then turned into a 27-acre lake. During the day he worked as district conservationist, and the rest of the time he raised fish and invested in land. Rising catfish prices and a turnaround in land values in the next few years brought him good returns. In 1987 Smelley made the leap into full-time catfish farming with 106 acres of ponds. He sold his fish every year to Southern Pride.

Smelley recognized opportunity and took it when he saw it. He had seen the catfish industry go through low points and had seen it recover. He had witnessed one processing plant in the county shut down and seen another one struggling. Convinced that an upturn was inevitable and that he knew enough to try his hand at processing, he decided after a year or two that it was a good time to enter the business himself.

He and four other farmers willing to invest with him began looking for funds. Finding capital for the building took longer than the construction. Banks did not want to take the risk after the recent plant failures in the area, and neither federal nor state money was available. Finally, an industrial bond was issued and an outside bank found that would help. Harvest Select Catfish Inc. purchased property in Uniontown, Alabama, and was processing fish by fall of 1991.

The effects of previous plant failures posed several obstacles. Area farmers were fearful they would not get paid for their fish, and customers were afraid to place orders for fear of not getting their desired product or not getting it on time. It worked, says Smelley, only because the partners "were committed. We didn't have a choice. There were no ifs, ands, and buts."

The most urgent difficulty for the new Harvest Select plant was the same one both Glovers had had—having enough money to run the plant day-to-day and week-to-week. "People never realize how much cash it takes to run an industry like this," says Smelley. "We opened our doors thinking half a million dollars cash was a lot of money. We spent that in a week! It was a rude awakening."

While generally less drastic than early on, unpredictable changes in demand, supply, and price are still the norm throughout the industry. "The market changes on you, and it's a daily struggle," Smelley says. In addition, costs often turn out to be much higher than expected. There were times when Harvest Select employees and bills were paid on time but the six stockholders did not pay themselves for their fish so that the plant could continue operating. "If we had not done that, we would not have stayed in business long," he says. Sales came more easily as the company developed a name for itself, but also partly because demand grew and the entire industry was maturing somewhat. Processors like Southern Pride and Harvest Select are still in business because their owners and managers are able to get the necessary resources and know-how to operate successfully.

16
Processing Matures

> This is not a normal business. You're buying something you can't see, and you're selling something that you haven't bought, and you've got to pay for it before you get paid.
>
> —George Smelley

Early catfish processors faced many daunting challenges, some of which they still deal with today. One such difficulty was the procuring of fish. Every day, processors needed to obtain enough fish of good quality and the right size to fill customers' orders. This was a very uncertain process. As independent growers, most catfish farmers naturally sought to improve their own circumstances. If they became dissatisfied with a processor for any reason, they would typically promise their fish to another plant if one was available. Richard True remembers his years managing different plants: "I would come home at night and get on the phone trying to line up fish for the next day. I never knew whether the farmer would sell them to me or not."

The price processors offered farmers for their fish was often a contentious issue. When processors could not obtain enough fish, competition for the available fish made it necessary for them to increase what they paid. When there were more fish than the processor could sell, however, farmers had to accept lower prices but were not happy about it. Joe Kyser Jr., whose father had been the first to raise catfish commercially for sale to STRAL, procured fish for Country Fresh. Sometimes farmers withheld their fish after hearing that prices were going to rise, and he would not be able to line up enough. Then, when prices did go up, there would suddenly be too many farmers trying to sell him their fish, and he would have to figure out how to buy from all of them so the plant would not fall from their favor. It was a constant juggling act.

Procurers needed to know how many fish a farmer had. Generally, farmers could tell whether they had forty thousand or a hundred thousand fish in their pond. Sometimes, however, even farmers' estimates were off, and processors ended up with either tens of thousands fewer or tens of thousands more than they needed. Knowing how to work with farmers' estimates became easier as working relationships developed, because "you kind of know the farmers after being around them so long," says Kyser.

The procurer's job brought with it other problems as well. Harvest scheduling was essential to getting the fish they needed, and schedules were disrupted when crews could not proceed as planned. If harvesting jobs early in the week took longer than predicted, farms scheduled later had to be put off. When off-flavor occurred, harvesting might be held up for a few weeks to several months. There were times when more than half the fish available were off-flavor. The delay was a strain on the plants, because they could not depend on having fish to harvest. Joe Kyser remembers one instance in which a farmer tried for five years to get one of his ponds back on flavor and never could.

Farmers often resisted when harvests were delayed or refused. Additional time, particularly in the heat of summer, meant additional feeding costs and health risks while maintaining the fish. Off-flavor was a disappointing and trying experience. "When a processor tells you your fish are off-flavor, you don't say anything," says Harry Miller. "You go home with your head tucked down. You don't argue, but it hurts."

~

Harvesting itself held its own set of challenges and specialized technology. At the beginning, crews consisted of the farmer, any employees he might have, and relatives who could pitch in. "It was kind of a family thing back then," Bill Kyser explains, and "the more the better, because it got to be pretty tough sometimes." The job was cumbersome, exhausting, time-consuming, and expensive.

Some early ponds were almost impossible to harvest, and labor costs were high. If the seines had to be hauled through water much deeper than six feet, they missed the fish, which went under the seine. Before the seine reel came into common use around 1990, crews had to unload the heavy seines from a trailer at pondside and load them back on, even heavier, after harvesting. The system called "topping off" advanced harvesting methods significantly. Rather than hand-selecting fish of marketable size, the grading sock automatically left those behind that were too small. Even after tractors and mechanical booms and winches came on the scene in the late 1960s and early 1970s, the fish still had to be dipped out of the water with nets and loaded into boom baskets by hand. Tall cranes or booms lifting large baskets of fish directly out of the water made it unnecessary to hand-fill buckets and pull them up and down the banks. Not only did the booms lift more fish out at a time, together with the grading sock they eliminated much of the handling stress the fish had previously undergone.

Harvesting old ponds with debris-strewn bottoms was uneconomical. Because processors and farmers lived in the same communities and in many cases had known each other a long time, crews continued to go to such ponds for a time, with the understanding that the arrangement would not continue indefinitely. Most farmers eventually accepted the reality of the situation and, if able, built levee ponds. The rest had to settle for more limited production and different markets, and many eventually stopped raising fish commercially altogether. Catfish farming gradually shifted its concentration to areas south of Greensboro, where the land was less hilly and shallower ponds with relatively flat bottoms could be built.

Easily seined ponds solved many of the industry's problems. Ponds could often be built next to each other, allowing time to be saved that would otherwise be spent loading and hauling seines and moving heavy vehicles from place to place. Ponds as large as twenty acres, if evenly fifteen feet or less deep, could still be seined with new, longer and wider seines. Seine reels were eventually built to hold two thousand feet of seine, and harvesters simply put the net out between two tractors on either side of one end of the pond and the tractors pulled it through the water to the other. Although one pull-through still did not catch all the fish and ponds still had to be seined more than once, the method was nevertheless more efficient and more economical.

Shallower ponds required much less water. The year's worth of collected rainwater and watershed drainage was not lost, because ponds did not have to be nearly empty to be harvested. Not having to power pumps to move huge volumes of water to refill ponds also helped save electricity. Processors were no longer flooded with surpluses of fish every fall, because ponds could be restocked immediately after harvesting and because topping off allowed them to be harvested and restocked throughout the year.

Muddy pond bottoms slowed harvesting considerably. The water-soaked bottom edge of the seine, the mud line, was weighted to drag on the pond bottom to keep fish from escaping underneath. As the seine was pulled through the water, the mud line would dig itself deeper and deeper into the mud until the seine would no longer budge. Joe Kyser got the idea of using a motorboat to push the top of the seine along. That helped keep the mud line from becoming bogged down in the mud so tractors could continue to pull the seine through. Although motorboats are now a common part of harvest equipment, mud can still affect the length of the job. "Sometimes you can seine a pond in one hour," says Kyser, "sometimes one in a whole day."

Early on, when hand-harvesting was the norm, moving around in the mud, especially when it was waist deep, was no easy task. For the harvest crews, such conditions could cause problems. "We used to have to go out and pick up what fish were left in the pond," Joe Kyser remembers. "You'd have to wade out. Some [workers] would be ankle-deep in mud, some waist-deep." Other men would have to reach and pull them out, or a boom basket was lowered to them and the winch would have to draw them out.

The mud did occasionally make for comic moments. Crew members might break up in laughter at the antics of a fellow worker trying to extract himself. Laughing in that situation could be risky, however: "*You'd* throw *them* out there if they laughed too hard at you," recalls Kyser.

The channel catfish's long, dorsal spine can pierce harvesters' boot soles. Together with a slippery, thick-muddy bottom, the spines can make harvesting a painful experience. "I stepped on a fish one time and picked my foot up," recalls Kyser. "I put it back down and got stuck again and fell down and took a spine in the palm of my hand."

The same mud that holds water in the ponds so effectively made the roads a sticky, slippery mess when it rained. Harvesting trucks regularly got stuck, and when farmers brought tractors over to help pull them out, the situation did not always improve, according to Kyser. "We got stuck having big tractors trying to pull you out, breaking chains, and getting more tractors stuck. And the plant would be hollering, wanting fish, and asking how come you couldn't get there. You make a mud hole, and it always gets worse instead of better." Topping the roads with gravel allows harvesting and maintenance rain or shine, but replacing the gravel is ongoing and expensive.

Just like mail carriers, harvest crews have to work in all types of weather, including icy cold and steaming hot. Winter was worse than summertime, according to Kyser. When ponds are covered with ice, holes have to be made at both ends to drop the seine in at one end and get it out again at the other. He remembers days in winter when the ice was so thick on the ponds that he had to climb out and walk around to the front of the boat, jump up and down to break the ice up, and then, risking falling in, walk a few feet farther to break up more ice so the boat could move forward and pull the seine the length of the pond under the ice.

On one occasion, a sleet and ice storm froze the roads. Somehow, "we caught the fish, got to the plant, and put the fish in the holding troughs,"

Kyser recalls. The fish were to be processed the next morning. But "it was so cold nobody could get to work, and we had to hold the fish two or three days. By the third day, the electricity went off because ice froze on the line, and we didn't have any power. After going to all that trouble, the fish died."

Keeping the fish alive is the greatest challenge in hot weather. "You can crowd fish in the wintertime and get by with it," notes Kyser, "but in the summertime that water gets hot, and they die on you. You've got to be on your toes to keep 'em alive. Fish can die on you in five minutes." Harvesting and transporting fish to the plant also takes much longer in the summer, because fewer can be held in the tanks. Whereas in the wintertime approximately twenty thousand could be hauled at a time, in the summertime the same tanks would safely carry only twelve thousand without causing too much stress.

~

The process at the plants became steadily more mechanized and automated. Shoveling and scooping fish onto conveyor belts soon became too slow for the increasing numbers coming in and did far too much damage to the fish. Today, hoists raise and unload baskets holding four and five thousand pounds at a time. With the advent of electronically operated machinery came a grading process done by computer. Another major development was the fileting machine. Until the mid-1980s, each fish was being fileted by hand with a knife.

Recognizing its potential with catfish, Townsend Engineering Company soon replaced the Model 66A skinning machine with an improved model that could be used in small processing operations and was more effective and simpler to use. The company also strongly recommended that machine operators wear gloves furnished with the newer Model 600 in order to avoid injury.[1]

The skinning machine revolutionized the catfish industry almost overnight. Within a year or two after Richard True and his partners made the first one work in 1967, several plants in Mississippi obtained their own. A few changes over the years made it more effective. More than thirty-five years later, the Townsend skinning machine is the primary machine used in the industry. Richard True maintains that getting it was the key to making catfish processing commercially feasible.

~

> "Off-flavor"—an "earthy-musty" type of flavor the fish apparently absorb from the pond's environment. The detection [and] prevention has been frustrating scientists for years.
>
> —Jimmy Carlisle

Throughout the industry's history, the off-flavor problem has been inconsistent and varied among ponds. Originally, some people thought it came from stirring up the bottom when draining the pond water. But even fish in ponds that had not been stirred up could have the same unpleasant odor and taste. Some thought the unpleasant flavor might have come from feeds. The problem begged for scientific investigation.

The earliest systematic study of the off-flavor problem in catfish was begun in 1970, when Tom Lovell and other biologists at Auburn University began collecting fish from processors and from some of the ponds in West Alabama that had been off-flavor. The researchers set up a panel of individuals to evaluate the flavor of the fish. The panel's members tasted hundreds of samples of fish that were allegedly not off-flavor and fish that allegedly were. With each one, they concluded either that it tasted fine or that it had the typical bad flavor. Next, they found a term to describe the flavor, settling on the description "muddy-musty." They still did not know the cause, but having a description would help them find it.

Determining statistically that off-flavor was a real phenomenon was important progress, because some farmers refused to admit it even existed. The Greensboro processors had difficulties with some farmers whose fish they refused because of off-flavor. Those farmers maintained that off-flavor was nothing more than an excuse, a ploy of the processor, who, they thought, for some reason did not want to buy their fish. The researchers met periodically with farmers, assuring them that off-flavor was only too real. Convincing them was a gradual process and very important in getting them to cooperate with the processors.

The next steps involved scientific detective work. The researchers had to find out what agent or conditions were responsible. During the early 1970s they visited farmers and asked about their feeds and feeding routines. "Which pond was your off-flavor in?" they asked. "How many fish were in there?" "Did you notice anything unusual about the pond?" "Did it have a lot of algae?" "Was it old?" "Was it brand new?" "How much were you feeding the fish?" and so on.

They found reports in the scientific literature of similar off-flavors in Israeli ponds and in Cincinnati water supplies, off-flavors that had been traced to algae that produced a musty-type compound. The researchers guessed the catfish ponds might be harboring substances similar to these and needed to determine whether the chemicals were getting into the catfish and whether they were actually the source of off-flavor. Through experiments, they found that the off-flavor compounds were water soluble and that the fish could absorb them directly from the water. They also found that putting off-flavor fish into clean water caused the off-flavor to disappear just as quickly as it had appeared. Apparently, even a severe case of off-flavor could theoretically be cleaned up within three to five days.

This first revelation that the off-flavor in commercially grown catfish was caused by an organism growing in the ponds came around 1974. Finding the specific types of algae that were causing the problem was more difficult to determine, because several types of algae are always present in any pond. Just because one type predominates did not necessarily mean that that one was the culprit. Later, the researchers would learn of other organisms that also produce off-flavor.

By 1981 scientists had found a connection between the intensity of off-flavor in pond fish and the feeding rate. They found that feeding sixty or more pounds per acre per day would lead to "intense off-flavor." By that time, several courses of action were available. Farmers could wait for a big rain to flush the ponds out. They could try applying low concentrations of copper sulfate, hydrated lime, or potassium permanganate, all thought at the time to possibly help. None of these measures worked consistently, however, and all were costly.

During the mid-1980s the researchers succeeded in pinning the source of the off-flavor they had found in Alabama on a species of algae called *Anabaena.* A taste panel actually detected a number of different off-flavors in the fish that the scientists were later able to identify chemically. One of the most common compounds was geosmin. A solution to the problem was not found as readily, however, especially because destroying the geosmin-producing algae could also eliminate part of the oxygen supply, produced by algae, that is so necessary for growing the fish. With the help of former AU graduates managing large ponds in Mississippi, they calculated how many fish from a pond would have to be sampled to be 95 percent certain that a decision of off-flavor was correct. This was a major step toward helping processors make reliable decisions on whether to harvest a pond or not.

An ounce of prevention is worth a pound of cure.

—Old Proverb

Processors agree that one of the primary requirements for retaining customers is maintaining a high-quality product. In the early years of the industry, the flavor of the fish was less of a concern. Most people who already ate catfish were accustomed to the slightly muddy flavor of the wild fish. As marketing and sales improved, however, more and more of the public were tasting farmed catfish for the first time, and sales suffered when off-flavor fish got into the stores. According to Joe Glover Sr., off-flavor has been a continuing nightmare in the catfish business. Quality in processed catfish today still overwhelmingly means being on-flavor. Processors are constantly on the alert to keep off-flavor fish out of the plant.

Originally, decisions to harvest and process a load of fish were based on their smell. "In the beginning," recalls Randy McCrory, "we'd take the fish, clean them, and smell the stomach cavity under hot water." If the fish smelled off-flavor, they were either rejected or, occasionally, harvested and held in tanks or ponds till they cleared. The sense of smell is subjective and imprecise, however, and could not by itself always accurately and reliably measure the presence of off-flavor.

Alabama's processors turned to AU scientists to develop an objective test. Researchers had identified two related compounds from off-flavor fish in West Alabama's ponds, one of which was methylisoborneol, or MIB. Using very complicated equipment, they found that minute concentrations of MIB, as low as 0.6 parts per billion, were enough to be perceived. Equipment to detect such low concentrations in the fish was too intricate, however, and not practical for use in a processing plant.

Having procedures for a human taste panel in the plants would be too time-consuming, so Joe Glover looked for individuals with very keen senses of smell and taste for the particular off-flavors Farm Fresh was encountering. During the mid-1970s he bought a microwave for cooking sample fish before they were processed. Randy McCrory was one of the employees Glover chose to do the tasting. McCrory had eaten river catfish as a child and had always found them delicious. Over the years, after eating the farm-raised fish, he had begun to notice the difference in flavor between the two. Under Glover's direction, McCrory began taste-testing Farm Fresh's fish. He became responsible for testing essentially every batch of fish that went through the plant.

Although taste tests provided a more effective way to control off-flavor, it was still the primary dilemma facing plants when Hormel purchased Farm Fresh in 1983. The quality control measures Joe Glover Sr. had initiated were tightened and extended by Hormel. As manager of Hormel's quality control process, McCrory gutted the fish and checked it for off-flavor, tasting the back, the tail section, and the underside, which was usually the slowest to return to good flavor. By 2003 most plants had an employee to whom a farmer brought in typically two fish for a taste check about two weeks before a scheduled harvest. The day before harvest, he brought in two more. If the fish still tasted good, the pond was harvested. Before the truck was unloaded at the plant, the fish were tasted one last time. That was, and still is, essentially the full-time responsibility of one individual like Randy McCrory. According to Joe Kyser Jr., McCrory could identify, by taste alone, which of about twenty different off-flavors a fish might have.

A lesser, but by no means negligible, problem for early processors was the question of what to do with the offal. Even now, with advances in breeding and processing, close to 40 percent of a catfish can still be unusable or unmarketable as food for humans.[2] For a plant processing roughly 140,000 pounds of catfish a week, that meant more than 11,000 pounds of offal a day. The ConAgra plant placed sealed containers of offal in the county landfill, where they were buried as deeply as possible. Yet, within three or four days, enough methane gas would be created in the containers that they would rise to the surface, where they eventually exploded and released a terrible odor.

Processors began to look at the potential of using offal in other products. Auburn researchers had found a very high nutrient content in fish meal made from catfish offal and encouraged them to develop some kind of industry to recover the product. Glover looked for feed mills and processors of meat by-products who would agree to turn the offal, high in protein, into something economically useful. He offered Farm Fresh's offal to a processor in Birmingham at no cost initially to have it made into sample cat food. After about a year, when the feed product proved successful, Farm Fresh began getting paid in return. ConAgra took its wastes from its Greensboro plant and other processors to its plant in Mississippi and processed them into fish meal. Now, tank trucks pick up offal every day from Alabama and Mississippi plants for use as fish meal in animal feeds.

> Anyone can build a plant. Anybody can buy equipment. It's how well you get people and motivate people that makes the difference.
>
> —George Smelley

Jobs on the processing line have not changed fundamentally, but the labor situation has. When the STRAL plant was built, work on the line was seasonal, depending on the availability of fish. One shift of workers did all the processing. Thirty people were working on the line in ConAgra's STRAL plant around 1970, processing between ten and twenty thousand pounds of fish a day. With year-round supply came year-round employment, and with increasing sales came more processing jobs. By 2003 more than 780 people were employed by Southern Pride and were processing more than ten times as many pounds as ConAgra had produced thirty years earlier.

Most existing job opportunities in catfish farming areas are in the plants. Modern plants employ a number of people in offices and shipping departments in addition to those on the processing lines. Harvest crews provide jobs for a few. Limited numbers find employment on farms.

The West Alabama region, known as the Black Belt because of its rich black soil, where the state's two largest plants are located, includes some of the nation's poorest counties. Unemployment among the African American population is high, and public assistance has long been a mainstay to many families there. Processing lines have opened hundreds of minimally skilled jobs over the years that have been filled virtually exclusively by blacks, and a large proportion of those are women. This is generally the case for plants in other states as well. Reluctant to lose their public assistance, many recipients who are able to get jobs in the plants try to manage their hours of employment to keep them within the required income limits. As a result, a significant number do not work the full forty-hour week. "If a person makes their limit at thirty-five, then they can't come back to work or they give up their free housing or food stamps," says George Smelley. "That's been a tough obstacle to overcome."

In the beginning, employee attendance was fairly consistent and reliable, but over the years, absenteeism has increased. Plant managers cannot be sure from week to week or even day to day how many employees they will have. They may expect all of them to show up on Monday morning and find only 70 percent there. Eighty percent might be there on Tuesday and 100 percent on Wednesday. When a plant does not have enough workers on a particular day, temporary substitutes can usually be found. Jobs on the line require only minimal training, and West Alabama harbors a large pool of unskilled labor. Many local residents have worked on the processing lines at one time or another.

One way of stemming some absenteeism has been to provide services in the plant for which employees would otherwise have to leave work. "Many times, employees would come up to their supervisor and say, 'I got to be off this afternoon 'cause I got to go by the welfare office.' That employee used to take half a day off to go to the welfare office or do other things," explains

Joe Glover Sr. Southern Pride arranged for a space in the plant and got permission from the agency handling welfare for a representative to be at the plant every day. Employees could then take care of their welfare business on their break or at lunchtime and not have to leave their jobs.

Long-term retention of line workers is another serious situation plant managers face. Employees are hired on a permanent basis, but most end up temporary. Average turnover in plants is between 100 and 200 percent each year, according to Richard True. Six hundred people may work there on a given day, but over a two-year period, twelve or fifteen hundred may have been employed.

The notion held earlier by some that a high incidence of poverty and unemployment would make the rural South fertile for almost unlimited industrial growth appears simplistic now. Transiency and unpredictable attendance are said to be a limiting factor for future growth in the industry throughout the South, not just in Alabama. "They're seeing it in the chicken industry and in certain parts of Mississippi," notes Jimmy Carlisle. "You eventually deplete your labor force. At one time or another everybody has worked there, or their mother has, or somebody."

The plants' ability to produce, and to do so efficiently suffers, according to Joe Glover Sr. Processors have developed "incentives that would recognize employees for their quality of work and for their attendance. We had to make every effort to show special attention to the ones that are regular." Incentive plans instituted by employers now include bonus checks, personal parking spaces, appreciation plaques, public recognition, and a rate of pay proportional to the rate of attendance. Special events like employee family picnics and Christmas parties are organized every year.

Throughout the industry's development both in Alabama and other catfish farming regions, African American residents have had little opportunity to enter commercial catfish farming or processing as either owners or operators. They tend to be poorly prepared to be either entrepreneurs or skilled employees in this fast-paced and nontraditional industry. They generally lack the financial resources, the information, and the training. The incentives put in place by plant owners and managers clearly seek to create in employees a sense of pride in their work, but a large turnover continues to occur primarily among younger employees between eighteen and twenty-five years of age. In the meantime, a history of impoverishment and racial tension, as well as other factors, creates an extremely complex problem that employers struggle to deal with. The catfish industry is a major contributor to Alabama's economy. If it is to continue to grow, the problems of attendance and turnover among processing line employees will have to be better understood and largely resolved.

Traditionally, industries have developed because of people's need or desire for a particular item or service. Some industries, especially since World War II, arose originally out of the technological ability to produce certain kinds of products. Catfish farming is perhaps among the latter, having come into being when few people even dreamed of eating pond-raised catfish. Markets for the fish had to be created, and building demand for a brand new product took a long time and a great deal of effort.

Having developed so quickly, the industry functioned at first without established controls or rules of conduct. Competition between processing companies desperate to establish their hold on markets became destructive. By 1971 processors in Alabama and elsewhere saw clearly that a more effective marketing system had to be developed. The industry was killing itself by competing for the limited demand. "Something had to be done. We were all going broke," recalls Richard True. "We were starved for sales. We had more processors and far more fish than we had sales, and we were selling by cutting each other's prices. We had gotten to the point that nobody was making money. So it was either get together and try to work out a solution where we could stay in business, or. . . . "

Leave the business altogether? No one wanted to do that. Processors generally recognized that in order to sell fish, competition would have to be stabilized somehow. Everyone agreed that marketing programs were best run by processors and not producers. In September 1971 the American Catfish Marketing Association was officially founded by processors in Mississippi who wanted to find a way to fund marketing activities adequately and permanently, in an organized manner. The national catfish marketing cooperative would provide a way to communicate and exchange market news such as prices and other information.[3] Recognizing that they were "all in the same boat," they tried together to improve sales. "We could do so much more that way," says True, who was instrumental in organizing the group.

Research was conducted, and ideas about effective promotion were developed by economists such as E. W. McCoy and others at AU and elsewhere. Their ideas informed the thinking and plans of the group. Promoting catfish generically rather than by brand was deemed most beneficial for the industry as a whole. Association members recognized that, in the long run, the more sales they promoted, the better things would be for all of them.

From the beginning, the organization took a consciously cooperative approach, in stark contrast to how its members had previously related. True worked to encourage negotiation among the parties, who had until then viewed each other as adversaries. "Before, there were seven people out there trying to sell to the same person," he recalls. "I felt it would be much better if we sat down and talked to one another. That was one of the big things—a conference call every Monday. If there was a problem between us, we probably could solve it." The organization also undertook to police itself. "If one processor was doing something we didn't think was right," says True, "we used the influence of other processors to straighten it out. They all agreed to abide by the majority. And it did help the industry."

Building markets was not a simple task. The channel catfish's public image was a significant obstacle. Originally, catfish was eaten by a relatively small proportion of Americans and as a food fish was simply unknown to most. In Alabama, it had been available only from people who fished the rivers and served only in a few restaurants located next to certain rivers. A negative attitude toward catfish was common, even in the South, among some groups who viewed the fish as inedible or undesirable because of its role as a scavenger and its association with the diet of poorer segments of the population.[4] Reports of poor quality in pond-raised catfish only hurt the fish's image more. Changing these attitudes required a concerted public relations campaign.

The fish's reputation turned around gradually. Widespread public education on nutrition and health benefits of eating fish paid off. Fish products became recognized as a good food for all consumers. Greater consistency in the supply and quality helped. Many people who thought they would not like it found they did like it after all. Customer orders began to outpace what the industry could supply. By 1975 catfish had "extremely wide acceptance" and was "priced as a gourmet product."[5]

In Alabama, Joe Glover Sr. and his sales staff continued to promote Farm Fresh catfish independently of the Catfish Marketing Association. During this time, Farm Fresh also added a portion-controlled, prebreaded product to its inventory of catfish filets and steaks. The new products were intended to appeal to working housewives looking for convenience. Demand increased for Farm Fresh's products, and new contracts kept coming in.

Glover had always tried to deal with competition and hard economic times by increasing plant efficiency and maintaining the company's reputation for quality. "We prided ourselves on turning out a quality product," he says. "When a competitor would come to our customers with a low-ball price and take that customer away from us, a great deal of the time we got that customer back." Joe Glover Jr. attributes his father's marketing success not only to his hard work and his record as a pioneer in the industry but also to his character. "My father never considered himself a great salesman," he says, "but when he called on a potential customer they sensed he was genuine and honest."

Long-distance and international sales were more exception than rule during the first years of the Alabama industry. The trend toward freezing the fish created additional possibilities for transport. By 2003 Southern Pride had a fleet of thirty to forty long-distance refrigerated trucks delivering fish in nearly every state in the nation. A few farms shipped food fish and fingerlings to certain European and Central American countries as early as the 1980s. In the late eighties big companies like Farm Fresh and Southern Pride began finding foreign buyers for processed fish. By 2000 some farmers shipped fish as far abroad as China. More recently, interest in catfish has grown in European countries as well as in Japan, where consumers enjoy the fish raw. The markets are still growing.

The biggest opposition to the Catfish Marketing Association came, from the very beginning, from among farmers. The farmers needed the processors but did not trust their having organized. While they liked receiving competing bids for their fish, they feared collusion among the bidders. They were afraid the processors would agree to hold down the price they had to pay for the farmers' fish. These fears contributed to the somewhat adversarial relationship that existed between the two groups.

Building trust took time and effort, but over the years a mutual understanding developed that the whole industry would collapse without organized marketing. "If we couldn't market the fish, farmers were going to be in just as bad a shape as we were," recalls True, "and they had to accept the fact that we needed to sit down together and not be adversaries, to get together and try to make the whole thing go." The understanding and cooperative effort that exist today in the Alabama industry took as long as ten to fifteen years to evolve. "It didn't start overnight. It was something we worked our way into," notes True.

In November 1986 the Alabama Catfish Producers informed Alabama farmers of the establishment of a new marketing agency, the American Catfish Institute. It had been founded by some Mississippi feed mills and financed by a checkoff program of about one-fifth of a cent per pound of feed sold to Mississippi farmers. The Catfish Institute, based in Belzoni, Mississippi, was going to conduct a broad campaign of generic advertising and improve the status of farm-raised catfish as a food fish.

Alabama's processors continued promoting sales of their fish independently, but an AU study prompted a change in the early 1990s when it showed that for every dollar the institute spent on advertising, a net return of thirteen dollars came back to producers.[6] About the same time, the statewide and the West Alabama producers' organizations decided to support the institute. They had approved a shift to generic advertising and urged Alabama feed mills to contribute to the effort. With the support of the two most influential producers' organizations in the state, the institute thus became the major marketing arm for Alabama's industry.

The Catfish Institute helped raise the status of farm-raised catfish nationwide. Today, the organization is part of the Catfish Farmers of America. Farmers are willing to participate because they realize that marketing is essential to the whole industry picture and because it is carried out as part of their organization, an organization created by farmers for farmers.

> It's been a learning process for us for thirty years. We're always finding better ways to handle and process fish.
>
> —Joe Glover Jr.

In the early 1990s John Jensen campaigned for a unique state-sponsored program of microbiological monitoring. In this program, catfish would be tested when they entered plants at the start of the line, at points along the lines, and at the end. Jensen argued that a label reading "Alabama Quality Assured" on plants' products would tell consumers that they were buying good-quality fish and that such a program would give the state's catfish industry a huge marketing advantage.

The farmers recognized the potential of the plan but approached it with some suspicion. The program would require them to work more closely and cooperatively with processors than ever before. Jensen convinced them it was a win-win situation: "It was based on selling more fish of better quality," he explains, "knowing everybody [could] take a deep sigh of relief that Alabama fish were safe. Nobody else was doing that." The farmers voted to help fund the program through the Alabama Catfish Producers.

Joe Glover Jr. also recognized the advantages of the program and offered to have Southern Pride share the cost for equipment, supplies, and a microbiologist to work out of the Fish Farming Center in Greensboro. Late in

1993 the Alabama Quality Assurance Program was established. Because no other state had any similar undertaking in their processing plants, the program placed Alabama's industry in a prime position to grow.

The Quality Assurance Program was instituted before the USDA's Food Safety and Inspection Service announced in the mid-1990s that steps were going to be required of the nation's seafood industry similar to those required of meat processors. The Hazard Analysis Critical Control Point (HACCP) system had been introduced during World War II. The system's processing principles were already used by the European Union and in Canada. In December 1997 the regulations began to apply to all U.S. seafood processors and handlers involved in interstate commerce. Before the Alabama program and the HACCP regulations, consumers eating catfish at restaurants could not be sure that the fish did not contain microorganisms that could cause illness or even death. "If you were a restaurant operator," notes George Smelley, "you could buy your fish from a guy who got in a boat, went down to the river, got fish, skinned them in his yard, threw them on ice, and brought them to you. You had no control over the environment they were processed in, whether they might have been on ice for three or four days, how they were handled, with whom they came into contact, *E. coli* . . . the whole nine yards." Anyone purchasing the fish to serve to others is now obligated to purchase only from HACCP-certified sources. Thanks to Alabama's Quality Assurance Program, the state was already a step ahead of the game (see fig. 22).

The speed and intensity with which the industry developed made for special problems for processors as well as producers. The repeated expansions and contractions had dramatic effects for the people involved. In 1969 three plants in Alabama were processing catfish commercially: Enfinger Brothers in Hartford, Fish Products in Suttle, and STRAL in Greensboro. A fourth, Southern Catfish Processors, was in the planning stages. In the Southeast as a whole, seven major plants had been established. That was more than the industry or the market could maintain. None broke even.[7]

Producers and processors did not yet have their act together. Fish crops were seasonal, and unsold inventories piled up every fall and winter. Processors were trying to figure out how to keep off-flavor fish out of the plants, and they could not fill restaurants' demands for controlled portions. Around 1970 one journal wrote of "approximately two million pounds of frozen product made up of a shotgun array of sizes and forms."[8]

Businesses appeared and faded from the scene. Small plants, especially, went out of business or changed hands about as fast as new ones were es-

Fig. 22. Quality control on the processing line, mid-1980s.
A section of the processing line at Southern Pride Catfish (now Southern Pride Catfish LLC/American Pride Seafoods). Alabama's Quality Assurance Program helped prepare the state's plants for the regulations of the USDA's Food Safety and Inspection Service in 1997. Here, before packaging or further processing, U.S. farm-raised catfish are inspected on the line in order to meet stringent quality requirements. (Photo courtesy of Southern Pride Catfish LLC.)

tablished. In 1980 more than twenty-five small-scale processors were operating in Alabama. Two years later, the number was half that. Smaller facilities faced pressures from all sides—supplies from farmers, access to markets, and competition from larger operations.[9] They suffered from competition with local fishermen selling wild fish and occasional competition from wholesalers selling fish they had gotten at low prices from larger plants. They had problems harvesting smaller farms with old-type ponds and getting quantities and sizes customers requested. State health standards for equipment placed an additional, relatively expensive burden on smaller operations.

The number of large processing plants also increased too rapidly for the supply and for the market, and most sat idle at times. Some closed down. Even by 1992, according to Randy Rhodes of Southern Pride, the thirty-five plants that existed then nationwide were, together, capable of handling two times more fish than were being produced.

Nevertheless, the overall trend was one of growth nationally and in Alabama. In 1974 more than 10 million pounds of catfish were processed nationwide, more than five times the amount five years earlier.[10] In 1982 close to 12 million pounds were processed in Alabama alone, bringing a sales income to Alabama's processors that was 13 percent of what all U.S. processors had made. By 1997 that figure had risen to 34 percent of the U.S. total.[11]

The first thirty years of processing offered a wild ride for those who decided to take the risk. Most of the "riders" were thrown off. Some held on. The bronco has become older, somewhat calmer, and is definitely better understood.

> We went through some frustrating times, but we made enough progress that we were confident we could work out the problems. We were determined to try to solve every problem that came up, and somehow or another we were able to make improvements.
>
> —Joe Glover Sr.

Notes

1. The Townsend Model 600 had (1) single-phase voltage for small processing operations, (2) a continuous tooth roll instead of teeth, (3) a stripper roll to keep the skin from wrapping around the tooth roll, (4) an optional air or water cleaning attachment, (5) a disposable razor blade-type of skinning blade of several widths for various depths of cuts, and (6) several other features to make the machine simple to operate.

2. True, interview.

3. "Catfish Processors Consider Exchange of Market Information." *Catfish Farmer* 2 (November 1970): 20–22, 24.

4. Cox, "Catfish Farming Industry of Southeastern United States," 109.

5. "Where Do We Go From Here?" *Commercial Fish Farmer* 1 (May–June 1975): 15.

6. Henry Kinnucan, Walter Zidack, and Upton Hatch, "Returns to Catfish Advertising and Optimal Spending Levels," Alabama Agricultural Experiment Station Bulletin No. 618 (1992).

7. Cox, "Catfish Farming Industry of Southeastern United States," 122.

8. Jack E. Greenfield, "The Catfish Industry—1971," *American Fish Farmer* 2 (March 1971): 13.

9. John Jensen and Howard Clonts, "The Catfish Industry," *Alabama Agribusiness* 22 (July 1983): 3; John Wilson Kilburn, "Catfish Farming in the Mid-South: Its Distribution and Potential Growth" (Ph.D. diss., University of Southern Mississippi, 1972), 46–47.

10. "NMFS Survey: 39 Million Pounds Harvested in 1971," *Catfish Farmer* 5 (July 1973): 28.

11. Jensen and Clonts, "The Catfish Industry," 2–3; "Annual Report: Commodity Producer Services," *Neighbors* 21 (December 1996): 45; Anisa Sumlar, "Farm-Raised Catfish: A Bright Spot for Alabama Agriculture," *Neighbors* 22 (September 1997): 16.

17
Reflections

It was really a crude operation when we started, and now it's a science.
—Buddy Lavender

When catfish farming was still in its infancy, people thought raising the fish in ponds would make an interesting hobby, provide families with a food source, and at best provide a small source of additional income to the growers. Many were skeptical about the investment they saw early fish farmers make in building additional ponds. "Man, you're talking about spending a lot of money when you don't know whether you can sell those fish or not!" they said. Even if farming were to be worthwhile on a small scale, many doubted it would grow significantly. "It'll never be like the poultry business!" was heard in and around Greensboro during the early 1960s.

Residents of the Black Belt had watched their landscape change repeatedly with agricultural enterprises that failed to live up to their dreams, but this time they were in for a surprise. Even with its ups and downs, the Alabama industry has grown tremendously. From 1982 to 2003 the total number of acres used for production rose from around 8,200 to more than 25,000, and the average size of a farm, from 33 acres to 127 acres. From 1993 to 2003 Alabama's share of the number of U.S. food catfish sold has nearly doubled, from 12 percent to 23 percent, and its share of the value of food fish sold has risen from 11 percent to 21 percent.[1] Well over a thousand employees in the state get their livelihood now in plants and related businesses.

Yet the catfish industry is different from what some early economists expected. In the 1930s Pres. Franklin D. Roosevelt "visualized a fishpond on every farm large enough to grow a quantity of fish sufficient to feed a family."[2] The dream of aquaculture as a source of inexpensive, good-quality protein available to rural families in their own backyards still seemed possible and relevant past the 1950s. Some even thought that catfish farming might hold hope for a way out of poverty for many of the South's rural poor.

Things evolved differently. The structure of the industry changed. Only a small fraction of the people who tried raising the fish commercially are still in the business. Costs and access to financial resources and markets have limited the potential for backyard ponds and small producers. The need to be economically efficient requires large acreages and the ability to build ponds and levees to certain specifications. Markets for smaller commercial farmers are limited. They generally have to settle for local niche markets, operating on an entirely different level than larger producers and processors, who can sell nationally and internationally. Economically, catfish farming is very much like other forms of agriculture now. There are fewer newcomers to commercial fish farming these days. Even years ago, it was not something for which a young person could afford to borrow money to undertake. "If one didn't already have the land and have it paid for, one could not make a living out of it," according to Richard True. Catfish farming is big business now.

> We would all be on welfare if it wasn't for catfish, no doubt. This would absolutely be a desolate place.
>
> —Richard True

The economy in West Alabama has always been based on agriculture. After the role of cotton weakened, the area could not seem to establish a solid, broad, or long-lasting economic base. Nothing seemed able to draw wealth to the area or provide significant employment. Many farmers in Alabama and elsewhere got little income from their crops and were losing their land. Many were planting anywhere they could in hopes of eking a little more from the soil. Often, receiving government money to keep land out of production seemed the only option. Agriculture in the 1970s and '80s generally did not fare much better, according to Wayne Shell. Regions with viable alternatives, such as the Tennessee Valley, weathered the slump and came out at the other end without turning to fish farming. West Alabamians were ready to try almost anything when catfish came along. Many farmers in Alabama and, for that matter, Mississippi, saw nowhere else to turn and converted their land to catfish ponds.

The local impact has been huge. The fish industry, partly by default, became the most productive activity Hale County farmers ever had. Catfish farming's success made it possible for farmers to keep their land, to continue farming, and to maintain the rural nature of the area.

The industry brought other economic activity. Local residents and individuals in nearby communities moved into new economic niches. They sold, installed, and repaired equipment; did electrical work; moved dirt; built roads; or hauled materials. The vast majority of contractors in the area were in business to build ponds. Businesses opened up, and employment increased. Manufacturers and suppliers sprang up to fill farmers' and processors' needs. A catfish farm of three to four hundred acres could need fifteen or twenty tractors and thirty or forty electric aerators, not to mention the tractor mowers, feed trucks, and other large equipment needed for harvesting.[3] Utility and fuel companies also benefited. Communication, transportation, and other service providers have been established to make supplies more easily accessible. In the early years, farmers had to order by catalog or drive many miles to get a pair of waders, and there was no way to obtain twenty tons of salt or gravel. Now they can obtain waders with a trip to a local store, and a phone call brings a truckload of salt within a few hours.

Like other rural areas dependent on a single large industry, Hale County's economy has become specialized. Ups and downs in the industry determine the pulse of economic activity. Locals who sell items to farmers can tell from their receipts how the industry is faring. A Greensboro tire salesman, for instance, maintains he can tell what the price of fish is by how many tires he sells.[4]

With the growing significance of catfish to the state's economy, the uses and value of the land are changing. At first, only marginal land not suited for pasture or row cropping was used to raise fish. Premium land was used for pasture. Now people are building ponds on the best farmland they have or can buy. Land prices are rising steadily, with the determining factor being the ease with which the land can be converted into ponds. More and more Alabama land is bulldozed into ponds, at a rate slightly higher than the national average. The job of planners like Elvin Wright is different from what it once was. "The old thinking has changed now," Wright says. "I put whole places in ponds, from one side to the other." As Check Stephens says, "The catfish industry did change the whole complex of Hale County."

The industry still faces some significant and persistent problems. Depredation, disease, and water and environmental quality issues continue to limit farming potential. Among the challenges facing processors in the foreseeable future are the difficulty of financing the high costs of operation, continuous change within the industry, and constantly rising labor costs. Labor is a key issue that will affect future growth of the processing industry. With high employee turnover rates, plant operators find it difficult to maintain a workforce that can process the large volumes of fish. The situation deserves study toward a resolution beneficial to all parties. In the meantime, in order to gain access to new labor pools, processors like Southern Pride have opened satellite plants in new locations.

The old problem of imported catfish has reached unprecedented dimensions and implications. Although it was a cause of concern among U.S. catfish farmers and processors, the number of imports brought into the United States posed a relatively minor threat until the end of the twentieth century. In 2002 the number surged to more than 13 percent of the total domestically raised crop.[5] Species other than the channel catfish are marketed so as to compete with them. In 1997 more than 941,000 pounds entered U.S. markets, and by 2002 the amount exceeded 40 million pounds.[6] The increase is primarily in higher numbers of filets sent from a burgeoning Vietnamese industry.[7] U.S. processors and producers are extremely concerned because the Vietnamese fish are produced and processed under different economic circumstances and under fewer constraints than those in the United States and can therefore be sold at much lower prices. Unless tariffs, laws, or some other actions bring the imports under control, the outlook is that domestic processors, including those in Alabama, will likely have a more difficult time selling their products. The income farmers are receiving from their sales has already dropped dramatically, and producers depending on fish sales to pay off debts face being forced out of business.

Predatory birds continue to cause great losses for farmers. The economic effects of depredation are difficult to quantify exactly. Industry and USDA wildlife damage control services estimate that Mississippi, Alabama, and Louisiana farmers lose a total of $10 to $30 million per year to cormorants alone.[8] Other species cause additional losses. As long as fish farms exist, the ready food source available in the ponds will continue to foster local populations of the birds.

Observers generally agree that future growth will not be as dramatic as it was in the past. Some say the issue is becoming increasingly a matter of balancing development with environmental concerns. Growth will be defined to some extent by limited land and water resources. Alabama's industry relies primarily on runoff from rainfall, an inexpensive source of water that can, however, also be unreliable. The state's fish farmers are, therefore, keenly aware of the limitations of their water supply.

The supply of groundwater in West Alabama is much less plentiful than it used to be. Although the well water is cold and deep, farmers have tapped it for a significant number of ponds. In 1960 STRAL's original hatchery began with water pumped from an overflowing well. Since then, the water level in the well has dropped to a depth of close to fifty feet or more, according to Richard True. A few farmers have been finding wells going dry as neighbors with deeper ones continue to draw from them.

Everyone in agriculture is aware that as the state's population expands, water management will become more crucial and use will be more closely watched, possibly even controlled. The state has begun to ask groundwater users to monitor their use. The implications of an inadequate water supply for the future of population centers and of agriculture are staggering, especially for fish farming. Some observers believe water will be the number-one problem the catfish industry will face down the road.

Water quality is as much a concern as quantity. According to Claude Boyd, environmental issues will be the next problem farmers have to deal with—effluents and issues about where the ponds can be placed. All types of aquaculture are going to be scrutinized. Concern exists regarding the quality of the water coming in as well as the quality of the water going out of the ponds. Farmers are aware that problems with effluent pollution will have effects not only on the environment but also on their own operations.[9]

High population densities are a major factor in producing pollution. In the case of fish farming, the enrichment of pond water by waste products of fish and decomposition of unused food increases chances for parasite and disease infection. Farmers' use of therapeutic chemicals to treat these problems has created concern. The concern also exists that farmers, especially those without adequate technical training, tend to use higher dosages than necessary, causing undesirable environmental impacts.[10] Some believe these chemicals, including antibiotics, metals, sulfa preparations, and others, could pose problems in the future. In Alabama, water quality researchers and regulators have examined pond effluent discharges into the environment. They have found no serious problems as yet, but they continue to pay close attention to the situation.

The issue of confined animal feeding operations has broken into the world of aquaculture in recent years with regard to effluents. Catfish farms have been designated somewhat differently as confined animal production facilities, and federal and state environmental officials and the Alabama Catfish Producers have established guidelines for environmental management. Rather than institute a permitting system, as has often been done in cases of environmental releases of effluents, the groups have developed a system of voluntary best management practices that meet Clean Water Act requirements. If there is a problem or a complaint, these practices will serve as the standards against which the situation will be evaluated.

Effluent water is expensive to treat. One method of conserving water and maintaining its quality at the same time is by pumping to adjacent empty ponds and allowing it to go through a natural cleansing process. Dan Butterfield uses this method successfully on his farm. His relatively small ponds are located next to each other and can be connected with hoses and pumps. The water is effectively recirculated between ponds and holding areas. Serious questions exist, however, as to whether such a system could be embraced on the large farms that are the norm today. Large pumping systems would be expensive to operate and maintain and would put large ponds out of circulation for longer periods of time than small ones, eating into a farmer's ability to produce and, therefore, into his profit.

Disease loss within the industry is tremendous. Very few chemicals are approved for use in treating catfish or other food fish. According to farmers, the industry is raising fish without adequate drugs. "The main thing hindering the fish farmer at this time is that they've taken all of the medicines away from us to cure the diseases and parasites on our fish," says Buddy Lavender.

The development of new drugs is a problematic issue. Bacteria can become resistant to treatments, and by the time a substance is cleared and approved by the FDA, it can often be ineffective. Determining and approving a drug as safe to give fish consumed by humans takes not only time but also millions of dollars. Most pharmaceutical companies do not consider the investment justified for a potential market the size of the catfish industry. Getting approval of therapeutic agents for food fish is a high priority for farmers. Researchers and the industry are working with the FDA trying to get clearance on many, and a few pharmaceutical companies are beginning to take some interest.[11]

The approach traditionally taken by fish disease specialists and farmers to solving disease problems primarily involved examining the fish, identifying the responsible organisms, finding a chemical, and treating the fish.

The "seek, identify, and treat" approach was aimed at finding something objective with which researchers could deal, according to John Plumb. "We looked at the gills. We looked at the skin. We looked at the lesions," he says. "Do we find parasites? bacteria? a virus? If we could [identify] any of those three, we had something firm to grab onto."

The traditional method was only reasonably successful. It is now considered by many to have been somewhat naive. The industry has learned there is more to the problem than originally thought. Treating a pond for parasites or disease no longer seems simple. The response of individual ponds to similar treatments can be different. "I have never had two ponds that are exactly alike," says Dan Butterfield. "Each has its own chemical, biological, and physical parameters." Researchers have found the effectiveness of chemicals is limited if other conditions exist that do not favor the fish's survival. In addition, treatments have an effect on the effluent from ponds and a certain level of toxicity on and off the farm to fish, people, and the environment.

A new approach in the struggle against disease has evolved in recent years, partly as a result of difficulties encountered with known treatments and the lack of acceptable substitutes. The approach involves "viewing disease in the context of the whole pond environment," explains Plumb. It is more holistic and total system oriented. It requires looking at a multitude of problems on the farm—including problems with water quality, nutrition, management methods, and parasites—and examining the circumstances or phenomena associated with these problems. This method has led to new ways of managing and handling the fish and promises beneficial impacts on water quality as well. The approach shifts some of the emphasis in fighting disease to management and prevention. New kinds of questions are being asked: "Why are these bacteria here now?" "Why has this parasite become a problem?" "Why is the virus killing this fish now and was not before?" Principles from the animal health industry have been adapted to fish health management. They involve factors affecting the health of the fish, which, in turn, influences their disease susceptibility. In aquaculture, some of these principles are related to feeding and stocking rates, which impact water quality and the conditions under which fish are growing.

The development of vaccines as preventive measures is a relatively recent effort. Researchers have been working on vaccines for catfish for about a decade. One or two have had moderate success, and a vaccine for enteric septicemia has been produced which initially appears to be effective. On the whole, much work remains to be done in this area. But even the best of vaccines will not prevent a problem if environmental conditions are not suitable for it to work or for the fish to survive. Researchers have learned that environmental problems can override almost any drug, chemical, or vaccine. As a result, none of them is foolproof.

Another potential front on which disease problems are being fought is genetics. The work being done uses gene alteration and traditional breeding techniques to develop fish that are less susceptible to disease or better able to fight infection. Genetic improvement of channel catfish has several additional goals, according to AU genetics researcher Rex Dunham. These include development of a fish that will grow to marketable size in a shorter time by converting feed faster and more efficiently, tolerate low oxygen conditions, and be easier to harvest.

What does this mean for the catfish farmer? For the foreseeable future, vaccines and genetic improvement remain areas primarily of great potential. The industry's approach to disease is going through a gradual transition to focus more on preventive health maintenance. Farmers can reduce the potential for disease in their ponds and lessen its severity when it does occur through proper management techniques such as maintaining water quality and stocking densities and health maintenance procedures. The approach has already shown positive results. "The industry as a whole has learned from these kinds of experiences that more does not necessarily mean better," says John Plumb. "This is one of the most important involvements of the catfish industry over the past thirty years."

~

Alabama's catfish industry has some unique qualities. For one, it is family-based. Alabama's farms are operated typically by the farmers rather than by hired managers. Farmers and family members do much of the work themselves. This has strengthened the Alabama industry because the farmers care about their work and about making it succeed. This motivation promotes development of an attitude, knowledge, and skills that make for good management. Good management is efficient management, and efficient management results in less cost. "We produce fish for less cost here. There's no doubt about it," says John Jensen. "Our farmers have been underpaid by five cents a pound and have survived and grown with that." Out-of-state investors considering entering the catfish industry look at the average return on investment over the past twenty years. In Alabama that return has been, on average, little better than ten to twelve percent. "Ten to twelve percent! Why do you do it?" they ask the farmers.

The answer to that question lies in the fact that the farmers want to continue living in a rural environment, on the land that, in many cases, has belonged to their family in a tradition of farming. In other aquacultural industries, most investors come from the outside and want to "make a killing" at them. Alabama's catfish farmers want to make a living at it. Consider, for example, what one West Alabama farmer expresses: "I loved the site where the ponds were built. I liked being down there. We were on a running creek with a little sand on it. I'd grown up in the prairie, and I was fascinated with the site. And I've always loved water, so I liked the idea of these vast areas of water. Back then, a ten-acre pond was a tremendous body of water, and I loved seeing that."[12]

Farmers working the land on which they live develop a knowledge of the land and a relationship with it, a feeling for the natural conditions. These help them farm more effectively. According to Jensen, an experienced fish farmer can probably walk out of the house in the morning, smell the air, and know how much his fish are going to eat. Living in rural areas, they also possess an independence and resourcefulness that comes from having dealt with and solved all kinds of problems on the farm from day to day. An outstanding example is Lenson Montz's development of the original paddle wheel.

Second, Alabama's industry is also community-based. Fish production, processing, and sales all take place generally in the same communities. The farmers tend to live within a close area and see each other frequently. Most appear to be active in their communities. Often, they are friends, go to church together, go to school functions together, play ball together, and watch their children play ball together. Members of the small farming communities generally know much of what their neighbors are doing. Researchers have found that word-of-mouth is a primary way in which ideas and information travel in agrarian communities, and when a farmer made money with his ponds or increased his acreage, the word spread quickly.[13]

Benefits occur when an industry is anchored solidly within a community. One such benefit is the informal resolution of conflicts and problems that might otherwise be disruptive and destructive to both the industry and the community. Years of associating with each other as peers in a variety of contexts tends to result in a certain level of trust. An event illustrating the advantage of such relationships is described with a chuckle by all the individuals in and around Greensboro who remember and were part of it.

After Nebraska Consolidated Mills bought STRAL Processing Company, it operated the plant as Country Fresh in the original STRAL building. Next to the building was a trailer in which the offices were located. The company's main competitor, Farm Fresh, established by Joe Glover, was operating right across the street. Whereas once Glover and Richard True had been business partners, they were now business competitors. They knew each other well, knew the community well, and were known in the community. Joe Glover Jr. worked at Farm Fresh at the time and recalls: "Richard True was operating the business just across the street from us—Country Fresh. Farm Fresh's loading dock was on the hill. When we backed our trucks up to the loading dock, the trucks were facing Country Fresh. One day, one of the drivers left the truck out of gear, and the brakes did not hold. The truck rolled down the hill. Richard True had a Cadillac his aunt had given him, and Clint Wilkinson, one of his partners—his pickup was parked in front of Richard's Cadillac. Farm Fresh's truck hit Richard's Cadillac and knocked the Cadillac into the pickup and the pickup into the Country Fresh trailer and knocked the trailer off the blocks. They accused Farm Fresh of deliberately trying to get rid of them." The accusation was made in jest. No one took serious offense. Instead of feuding or suing, the parties took appropriate responsible action and otherwise laughed it off.

> When there's a problem that needs to be solved, everybody gets together and solves it, because we all feel like we're a part of this thing. In other businesses, people tend to hold onto the information and don't share it with their neighbors—don't want other people to know about it. They become autonomous, and they don't work very well together.
>
> —John Jensen

A third and unusual key ingredient in the industry's success in Alabama has been the communication and teamwork that have become the norm among the people involved. When extension staff brought information to Auburn University faculty about real problems the farmers were experiencing, the researchers responded with practical knowledge the farmers could use. A working relationship evolved in which farmers consulted and learned from faculty and faculty learned from the farmers.

Alabama's farmers are a close-knit group and have generally shared interests and information. If one farmer has a problem and knows others in a similar situation, they typically talk about it among themselves. The seeds of this cooperation go back to the early days when there was very little infrastructure for fish farming and farmers needed each other just to keep their operations afloat. The prevailing feeling was one of cooperation, whether the task was building or repairing a piece of equipment, finding a way to haul fish, or something else. "There was no room for somebody to be particularly selfish and hoard information," notes one farmer. "There was no great reason to hoard information, because the markets were expanding and growing. We were competing against each other, but we didn't consider each other as competitors. We were jointly building an industry."[14]

The producers and processors in the state work together toward common goals. According to John Jensen, having everyone in an industry work together to solve problems, as they do in Alabama's catfish industry, is very unusual. That relationship did not come into being by itself or overnight, however. As an extensionist, Jensen focused to a large extent on finding ways to help the different parties avoid the conflicts that are so natural in that sort of relationship. Communication, trust, and understanding between the two groups have been easier to maintain because fish prices have remained up. The fact that the owners of the plants are fish farmers themselves and have a knowledge of farming helps also. At times, things can get problematic between the two, but good business practices and good communication have gone a long way toward helping the industry function smoothly.

Farmers know that they have to keep open lines of communication with the two large processing plants in the state in order to sell their fish. With energetic sales staffs and innovative management, according to Jimmy Carlisle, Alabama processors have been extremely successful in marketing farm-raised catfish nationwide. Because of this successful marketing, the processors have been able to expand the volume of fish going through their plants. This expansion has in turn assured farmers, for the most part, of a buyer for their fish.

Producers' and processors' organizations, formed with the specific intent of supporting the industry, have contributed to its functioning and growth. A positive fact of the farmers' organizations in Alabama is that they have learned to manage their industry themselves through cooperation and self-regulation. The Alabama Farmer's Federation, of which the Alabama Catfish Producers is a part, lobbies for and supports farmers as a group statewide. Regional associations have contributed by identifying priorities, deciding which problems to deal with, and trying to solve them. Because the industry is unevenly distributed in the state and farmers in the different areas experience different constraints, it seems reasonable that local organizations might best fill some local needs. The Catfish Farmers of America supports the industry politically on a national level and works with federal agencies on industry problems in ways that state and local organizations would be unable to do.

> It gets down to individuals and entrepreneurship, initiative and motivation.
> —Jimmy Carlisle

No amount of natural resources, research, extension, or other infrastructure would have succeeded in building the industry without people with the qualities that helped them see their way and steer others through the maze of obstacles that kept coming up. The progress of the industry is in no small measure a result of the contributions of many individuals. Two of them left particularly outstanding legacies.

Homer Swingle was one of those. As the founder and first head of the fisheries program at AU and as a top-notch scientist, he made a strong and lasting impression on the people who worked for and learned from him. He encouraged his faculty and students to develop new ideas, and once convinced of an idea's value, he always found a way to support it financially. He developed many fundamental concepts about population dynamics and pond aquaculture at a time when no fisheries existed yet in the eastern United States. His findings on commercial catfish production became the bible for the industry, which then built on the foundations he and the faculty he hired established. He impacted the industry also through the work ethic and habits he taught his students, many of whom later moved into jobs that figured importantly in the industry's progress.

Of the many people who had a direct hand in the industry's development in the state, Joe Glover Sr. stands head and shoulders above the rest (see fig. 23). Called the "father of catfish" and the "grandfather of the Alabama processing industry," Glover was on a mission from the very beginning. He wanted to see Alabama's catfish industry grow and succeed. "There were times he would like to have just thrown up his hands," remembers his son, "when he didn't have fish, when he didn't have labor, when he couldn't sell the fish, when there was no money. It would have been easy for him to just give up, but he didn't."

Glover won a special loyalty from everyone who worked with him. His consistent use of "we" instead of "I" reflected the consideration he showed them. Rather than tell people to go do a task, he would ask them to help him do it. If a task did not produce the desired results, he would ask, "What do we need to do next?" He made a practice of getting employees' feedback, ideas, and advice. Rather than claim that he had been responsible for mak-

Fig. 23. Joe Glover Sr. at his first pond, 1998.
Joe Glover Sr. stands next to his original catfish pond, the first pond harvested by STRAL Processing Company.

ing Farm Fresh so successful, he has said, "I didn't build Farm Fresh and make it successful by myself. I had some good help. Those employees were loyal. . . . In turn, I showed them every consideration that I could. . . . I was not the only one that was able to make it successful, but I feel a certain amount of pride and accomplishment for having been one of the early people in the business and able to watch it grow. I am just proud to have had a part in a young and thriving industry that has developed from a very small scale."

His vision is now a reality.

Notes

1. National Agricultural Statistical Service, "Catfish Production," U.S. Department of Agriculture, Washington, D.C., 1981, 1982, 1987–2004.

2. "New Farming Technologies Make Roosevelt's Dream Come True," *American Fish Farmer* 1 (December 1969): 17.

3. Joe Glover Sr., interview by author, tape recording, 27 February 1997, Special Collections and Archives, Auburn University, Auburn, Ala.

4. Kyser, interview.

5. Import data of the National Marine Fisheries Service are obtained from the U.S. Bureau of Census, Foreign Trade Division.

6. "Commerce Department Rules Vietnam 'Dumping' Fish on the U.S. Market," *Catfish Journal*, http://www.catfishjournal.com/id19.htm (accessed February 2003).

7. Jerry Crews, "Aquacultural Situation and Outlook," Alabama Cooperative Extension System, Auburn University, Ala., September 2002, Special Collections and Archives, Auburn University, Auburn, Ala., 4.

8. National Aquaculture Association, 13 July 1998, "Environmental Stewardship," "Depredation," http://www.nationalaquaculture.org/pages/depredation.html (accessed 6 October 2004).

9. John E. Bardach, "Aquaculture, Pollution, and Biodiversity," in *Sustainable Aquaculture*, ed. John E. Bardach (New York: John Wiley and Sons, 1997), 89, 91.

10. Ibid., 90.

11. Jimmy Carlisle, interview by author, tape recording, 13 January 1998, Special Collections and Archives, Auburn University, Auburn, Ala.

12. Kyser, interview.

13. Kilburn, "Catfish Farming in the Mid-South," 44.

14. Kyser, interview.

18
Postscript

A number of significant events in the development of the Alabama catfish industry took place after this project was under way. Among them, unfortunately, was the passing of Joe Glover Sr. Harvest Select Catfish has a new president, but George Smelley continues to be a stockholder in the company. In January 2003 Southern Pride Processing Company was acquired by American Seafoods Group of Seattle, Washington.[1] Joe Glover Jr. continued to head the company as president for a time but later resigned, limiting his involvement in the industry to the farming of fish on his own ponds. Randy Rhodes continues to oversee Southern Pride's sales and marketing activities. In late 2002 Southern Pride, the largest domestic catfish processing operation in the United States, employed more than nine hundred people and had annual sales in excess of $100 million.[2] The reach of Southern Pride products is expected to grow even further, now that it is aided by the marketing capabilities of American Seafoods.

The Alabama processing industry is still in part locally owned, but it has undergone some vertical integration. Since 1999 Harvest Select has been operating a feed mill in Uniontown. Until the sale of Southern Pride, ownership and decision-making in the industry was entirely in the hands of local entrepreneurs. For the first time in its extraordinary history, a significant part of the U.S. catfish industry is under the umbrella of a multinational entity. The balance that will emerge between multinational corporate ownership and local self-determination, and the direction in which that balance and vertical integration may take the state's catfish industry and its economy, will be for future observers to note.

Participants in the nation's catfish industry have achieved a good level of understanding of how the industry works, of its development, and of the conditions and constraints under which it functions. Yet, while the economic roller coaster may have smoothed out somewhat, problems continue to challenge Alabama's farmers and processors. The industry is undoubtedly here to stay, and its story continues.

Notes

1. Fish Information and Services, January 2003, search for "Southern Pride Catfish"; check "World News" at bottom of right-hand column, http://fis.com/fis/search/ (accessed 27 February 2003).

2. Southern Pride Catfish Company, "Southern Pride Catfish Acquired by American Seafoods," 21 November 2002, select: "News," http://www.southernpride.net/ (accessed May 2005).; Southern Pride Catfish Company, "Southern Pride Catfish Prepares for Growth and Expansion," 22 November 2002, http://www.southernpride.net/southernpride/news/American%20Seafood%20acquisition.htm (accessed 27 February 2003; page now discontinued).

Appendix: Trends and Major Events Affecting Alabama Catfish Industry Development

Items are listed in general chronological order within decades.

1880s and 1890s

- Channel catfish culture and reproductive behavior observed by federal researchers in Kansas and Washington, D.C.

1910s

- Manual on pond fish culture produced by Kansas Department of Game and Fish.
- Channel catfish successfully spawned by federal researchers in Iowa.

1920s

- Catfish propagated and hatching technologies developed at state fish hatchery in Pratt, Kans.

1930s

- Commercial bait minnow production begun in Southeast.
- Dam collapse investigated by Homer Swingle at Auburn University.
- Federal fish hatchery constructed in Marion, Ala.
- Farm Ponds Project initiated at Auburn.
- Soil Conservation Service (SCS) established.
- Findings on stocking, fertilization, and weed control reported by Homer Swingle and E. V. Smith at Auburn.

1940s

- Findings on spawning, rearing, and diseases of channel catfish reported by fish culturists in various states.
- SCS biologists' training and student fisheries curriculum initiated at Auburn.
- Pond-building promotion and assistance in planning and design begun by SCS.
- Groundbreaking advances in stimulation and control of spawning of catfish made by Kermit Sneed and Howard Clemens with the FWS at the University of Oklahoma.
- Catfish stocking begun by pond owners and reservoir managers.

1950s

- Search for a commercially viable food fish initiated by Homer Swingle.
- First work on parasites of warmwater fishes begun at Auburn.
- Work on nutrition and fish growth leads to development of Auburn No. 1 fish feed.
- Homer Swingle publishes work showing channel catfish can be produced commercially and economically.
- Investigations of management problems involved in raising catfish for food begun at Auburn.
- Catfish added to the research agenda at Marion.

1960s

- STRAL Company catfish hatchery established by Check Stephens, Richard True, and Bryant Allen.
- Supplemental vitamins added to Auburn No. 2, the first concentrated and pelleted fish food.
- Southeastern Cooperative Fish Parasite and Disease Project established at Auburn.
- Commercial feeds formulated for catfish developed by Western Grain and Purina Specialty Chows.
- Floating catfish feed developed by Ralph Wells Company.
- STRAL Processing Company established by Check Stephens, Richard True, and Joe Glover Sr.
- Model 66A Townsend skinning machine first used on catfish.

- STRAL processing plant established in Greensboro.
- Catfish Farmers of America founded.
- People with Peace Corps experience hired in AU fisheries department.
- STRAL Processing Company sold to Nebraska Consolidated Mills.
- Research on cage production reported by Rudy Schmittou at Auburn.

1970s

- Use of Terramycin and certain sulfa drugs in food fish receives final approval from the FDA.
- First systematic investigation of off-flavor in channel catfish conducted by Tom Lovell at Auburn.
- Farm Fresh Catfish Company established by Joe Glover Sr.
- American Catfish Marketing Association founded.
- Alabama Catfish Producers founded.
- FDA begins clampdown on use of some chemicals for fighting catfish parasites and diseases.
- Findings on the nature of channel catfish virus, broken back syndrome, and other diseases reported by Auburn researchers and others.
- Use of mechanical booms and winches in harvesting begun.
- Grading sock and "topping off" harvesting method developed by Paul Smith and Wayne McLendon.
- Catfish Division and State Catfish Committee added to Alabama Farm Bureau.
- Catfish fileting initiated at Farm Fresh.
- Consumer Marketing Labeling Act passed.
- Alabama processors increasingly refuse to harvest older-type ponds.
- Taste-testing procedures initiated at Farm Fresh.
- First paddle wheel aerator developed for ponds.
- Jimmy Carlisle hired at Alabama Farm Bureau.
- John Jensen hired as cooperative extension specialist in aquaculture.

1980s

- National Aquaculture Act of 1980 passed and National Aquaculture Plan adopted.
- Industry newsletter established and statewide trade shows and conferences organized.
- *Alabama Catfish Producer* begins publication.

- Fingerlings and food fish from farms exported to Europe and Central America.
- Connection found between intensity of off-flavor in pond fish and feeding rate.
- Alabama Fish Farming Center established.
- Antipoaching bill passed.
- Farm Fresh's processing business sold to Hormel.
- Amended version of "prompt pay" legislation passed.
- West Alabama Catfish Producers Association founded.
- Efficiency of electric paddle wheel maximized by Claude Boyd.
- Source of musty-muddy off-flavor in Alabama catfish identified by Tom Lovell and Auburn researchers.
- Southern Pride Processing Company established.
- Processed fish exported by large companies.
- Piedmont cage-raised catfish project initiated.
- American Catfish Institute established.
- Checkoff program approved.
- Exemption to fishing license requirements at fee-fishing enterprises passed.

1990s

- Piedmont Association of Caged Fish Producers founded.
- Harvest Select Catfish Inc. established.
- Support in Alabama industry for the Catfish Institute initiated and shift to generic advertising begun.
- New, larger Fish Farming Center constructed.
- Southern Pride operations expanded to Demopolis.
- Alabama Quality Assurance Program established.
- Funding for Fish Farming Center increased.
- Federal legislation passed to allow farmers to use certain methods to control predatory birds.
- Hazard Analysis Critical Control Point system regulations applied to all U.S. seafood processors and handlers involved in interstate commerce.
- Harvest Select feed mill established.

2000s

- Southern Pride Catfish Company sold to American Seafoods Group of Seattle, Wash.

Bibliography

Alabama Aquaculture Plan. A Project of Alabama Catfish Producers, Division of Alabama Farmers Federation, Alabama Department of Economic and Community Affairs, Auburn University, and U.S. Soil Conservation Service. Auburn University, 1992.

The Alabama Catfish Producer. Alabama Farm Bureau Federation, Catfish Division, Montgomery, Ala., 1986–97.

Alabama Farm Bureau News. Alabama Farm Bureau Federation, Cullman, Ala., 1979–87.

Alabama Farm Bureau Newsletter. Alabama Farm Bureau Federation, Cullman, Ala., 1993–97.

"Annual Report: Commodity Producer Services." *Neighbors* 21 (December 1996): 44–52.

"Aquaculture Plan Identifies Major Issues." *Neighbors* 18 (April 1992): 10.

Auburn University Department of Fisheries and Allied Aquacultures. Annual Reports. Records, 1934–97. Special Collections and Archives, Auburn University, Auburn, Ala.

Avery, Richard M. Notes on the History and Accomplishments of the Alabama Fish Farming Center and Status of Fund-Raising Efforts for a New Facility, no date. Special Collections and Archives, Auburn University, Auburn, Ala.

Ayers, James W. "The Catfish Market: Problems and Promise—Marketing Problems Demand Production Efficiency and Sales Promotion." *American Fish Farmer and World Aquaculture News* 2 (March 1971).

Bardach, John E. "Aquaculture, Pollution, and Biodiversity." In *Sustainable Aquaculture,* ed. John E. Bardach, 87–99. New York: John Wiley and Sons, 1997.

Bartonek, Frank A. "Catfish Farming." *American Fish Farmer and World Aquaculture News* 3 (July 1972): 4–7.

Branham, Lowell. "Catfish Farming." *American Fishes and U.S. Trout News* 16 (July–August 1971): 16–21.

Burdette, Ron. "Piedmont Catfish Industry, Coosa Valley RC&D, Project Measure No. 474." Oxford, Ala., 1985–91. Copy in author's possession.

Burnett, Jay. "After Church's." *Neighbors* 11 (March 1986): 10.

———. "No Way to Go but Up." *Alabama Farm Bureau News* 61 (spring 1985): 6–8.

———. "Something Fishy's Going On." *Neighbors* 7 (April 1982): 6–7.

Callahan, Ramon. "There's Gold in the Catfish Country." *American Fishes and U.S. Trout News* 13 (September–October 1968): 20–21.

Carlisle, Jimmy. "Commodity Digest: Catfish." *Neighbors* 6 (April 1981): 14–15.

———. "Commodity Digest: Catfish." *Neighbors* 6 (June 1981): 12–13.

———. "Commodity Digest: Catfish." *Neighbors* 6 (November 1981).

———. "Commodity Digest: Catfish." *Neighbors* 7 (April 1982): 10–11.

———. "Commodity Digest: Catfish." *Neighbors* 8 (January 1983): 10–11.

———. "Commodity Digest: Catfish." *Neighbors* 8 (April 1983): 10–11.

———. "Commodity Digest: Catfish." *Neighbors* 8 (June 1983): 21.

———. "Commodity Digest: Catfish." *Neighbors* 8 (August 1983): 17.

———. "Commodity Digest." *Neighbors* 9 (April 1984): 16.

———. "U.S. Catfish Industry as a Model for Development of International Aquaculture." Paper Presented at the Fifth Central American Symposium for Aquaculture (5th Simposio Centroamericano de Acuacultura), San Pedro Sula, Honduras, August 18–20,1999.

"Catfish Farmers Take Measures to Correct Problems." *American Fishes and U.S. Trout News* 14 (May–June 1970).

"Catfish Farming—Mississippi Style." *Catfish Farmer* 2 (May–June 1970): 23–25.

"Catfish Farming Saluted." *Catfish Farmer* 3 (January 1971): 42.

The Catfish Institute. "Aquaculture Provides Answer to Nation's Diminishing Seafood Supply." "Aquatic Network: News Archive." 22 June 2002. http://www.aquanet.com/features/catfish_institute/catfish.htm. (accessed 10 November 2003; page now discontinued).

"Catfish Processing Plants Increasing." *American Fish Farmer* 1 (December 1969): 14.

"Catfish Processors Consider Exchange of Market Information." *Catfish Farmer* 2 (November 1970).

"Catfish Research Could Flounder." *Neighbors* 11 (April 1986): 19.

"CFA to Hold Catfish Fry for Members of Congress." *Catfish Farmer* 3 (May 1971): 5.

"Commerce Department Rules Vietnam 'Dumping' Fish on the U.S. Market." *Catfish Journal.* http://www.catfishjournal.com/id19.htm (accessed February 2003).

"Conversations." Notes from Interviews with Alabama Fish Farming Center Staff, Greensboro, Ala., 11 June 1992. Photocopy. Special Collections and Archives, Auburn University, Ala.

Cox, Gary C. "The Catfish Farming Industry of Southeastern United States: A Geographic Inquiry into the Patterns and Processes of Regional Aquaculture." Ph.D. diss., University of Northern Colorado, 1972.

Creamer, Jamie. "Catfish Farmers Support Generic Ad Campaign." *Neighbors* 18 (April 1993): 6.

———. "Low Prices Pinch Catfish Producers." *Neighbors* 16 (December 1991): 4.

———. "New Catfish Feed Mill." *Neighbors* 16 (December 1991): 4.

———. "On the Rural Route: A Catfish in Every Pot." *Neighbors* 18 (February 1993): 4.

Crews, Jerry. "Aquacultural Situation and Outlook." Alabama Cooperative Extension System, Auburn University, Ala., September 2002. Special Collections and Archives, Auburn University, Auburn, Ala.

Cross, Frank B., comp. "Pondfish Culture in Kansas: A Selective Chronology." 1991. Special Collections and Archives, Auburn University, Auburn, Ala.

Doze, J. B. "The Barbed Trout of Kansas." *Transactions of the American Fisheries Society* 55 (August 1925): 167–83.

"Dr. John W. Jensen Relates Catfish Industry History." *AFC Farming News* 26 (December 1997): 19.

"Fingerlings: Three Veteran Producers Talk about the Past, Present, and Future." *Catfish Farmer* 4 (March 1972): 16–17.

"Fish Farming Center to See Funding Hike." *Neighbors* 21 (August 1996): 16.

"Fish Farming Today." *American Fish Farmer* 1 (December 1969): 11.

Fish Information and Services. January 2003. Search for "Southern Pride Catfish." Check "World News" at bottom of right-hand column. http://fis.com/fis/search/ (accessed 27 February 2003).

"Fish Is Big Business in Alabama." *Aquaculture Magazine* 25 (March–April 1999): 103–4.

"Food." *Time*, 16 May 1969, 112.

Greenfield, Jack E. "The Catfish Industry—1971." *American Fish Farmer* 2 (March 1971).

Griffin, Charles. "Catfish as Collateral." Address at the Second Annual Convention of Catfish Farmers of America. *Catfish Farmer* 2 (March–April 1970).

Hickling, C. F. "The Cultivation of *Tilapia*." *Scientific American* 208 (May 1963): 143–52.

"History of Fisheries at Auburn University." Notes for a Speech Given on the Occasion of Signing of Grant for the International Center of Aquaculture, 26 June 1970, Department of Fisheries and Allied Aquacultures. Acc. no. 97-29, B4C3, Box 3. Special Collections and Archives, Auburn University, Auburn, Ala.

Holley, Debra. "Oxygen Depletion Problem Faced by Catfish Farmers." *Neighbors* 1 (spring 1976): 21–22.

Jensen, John. "Alabama Harvests Catfish." *Neighbors* 5 (March 1980): 26–27.

Jensen, John, and Howard Clonts. "The Catfish Industry." *Alabama Agribusiness* 22 (July 1983): 2–3.

Jones, Walter G. "Economics and Marketing Considerations of Burgeoning Catfish Industry." *American Fish Farmer* 1 (January 1970): 20.

Kaufman, Kenn. *Birds of North America.* New York: Houghton Mifflin, 2000.

Kelly, Dave. "The Alabama Fish Farming Center." *Alabama Agribusiness* 22 (July 1983): 4.

Kennamer, Earl R. "Catfish Production and Marketing." *Alabama Agribusiness* 8 (April 1970): 4.

Kilburn, John Wilson. "Catfish Farming in the Mid-South: Its Distribution and Potential Growth." Ph.D. diss., University of Southern Mississippi, 1972.

Kinnucan, Henry, Walter Zidack, and Upton Hatch. "Returns to Catfish Advertising and Optimal Spending Levels." Alabama Agricultural Experiment Station Bulletin No. 618, 1992.

Kirby, Martin. "Catfish vs. Poverty?" *American Fish Farmer and World Aquaculture News* 4 (December 1972): 8–9.

Lawrence, John M. "Construction of Farm Fish Ponds." Alabama Agricultural Experiment Station Circular No. 95. Auburn University, 1949.

Ling, Shao-Wen. *Aquaculture in Southeast Asia: A Historical Overview.* Seattle: Distributed by University of Washington Press, 1977.

Madewell, Carl E., and Ralph J. Ballew. "Historical Development of Catfish Farming." *American Fish Farmer and World Aquaculture News* 3 (February 1972): 8–11.

"Market Outlook—A Processor's View." *Net Gains: The Newsletter for Alabama Fish Farmers* 1 (September 1982): 3, and 1 (November 1982): 2–3.

McLendon, Wayne. Interview with author. Auburn, Ala., 5 April 2000.

Montz, Ann. "A Note from One of the Wives Who Lived through This Story." 1996. Special Collections and Archives, Auburn University, Auburn, Ala.

Montz, Lenson. "Looking Back—1996." 1996. Special Collections and Archives, Auburn University, Auburn, Ala.

National Agricultural Statistical Service. "Catfish Production." U.S. Department of Agriculture, Washington, D.C., 1981, 1982, 1987–2004.

National Aquaculture Association. 13 July 1998. "Environmental Stewardship." "Depredation." http://www.nationalaquaculture.org/pages/depredation.html (accessed 6 October 2004).

National Marine Fisheries Service. Fisheries Statistics and Economics Division, Office of Science and Technology, National Oceanic and Atmospheric Administration. 11 March 2003. Select: "Imports," "January 1989–January 2002," "Catfish," "Table." http://www.st.nmfs.gov/st1/trade/cumulative_data/TradeDataProduct.html (accessed 20 March 2003).

Net Gains: The Newsletter for Alabama Fish Farmers. Auburn University, Alabama Cooperative Extension Service, and Alabama Farm Bureau Federation Catfish Division, May 1982–March 1983.

"New Farming Technologies Make Roosevelt's Dream Come True." *American Fish Farmer* 1 (December 1969): 17.

"News Briefs—Competition May Trim Farm Profits." *American Fish Farmer* 1 (February 1970): 2.

"The News-Creel." *American Fishes and U.S. Trout News* 15 (May–June 1971): 14.

"NMFS Survey: 39 Million Pounds Harvested in 1971." *Catfish Farmer* 5 (July 1973): 28.

Parker, Nick C. "History, Status, and Future of Aquaculture in the United States," *CRC Critical Reviews in Aquatic Sciences* 1(1) (1989): 97–109. Citing J. T. Bowen, "A History of Fish Culture as Related to the Development of Fishery Programs." In *A Century of Fisheries in North America*, ed. N. G. Benson. Special Publication No. 7, American Fisheries Society, Washington, D.C., 1970.

Prewitt, Roy. "Rambling Along." *American Fish Farmer* 1 (December 1969).

———. "Rambling Along." *American Fish Farmer* 1 (March 1970): 23–24.

"Production and Marketing of Farm Raised Catfish in Southeast Alabama." Alabama Cooperative Extension Service Circular R-20. Auburn University, n.d.

"Program Will Ensure Safe Catfish Supply." *Neighbors* 18 (October 1993): 19.

"Put Catfish Offal to Work for You." *Fish Farming Industries* 4 (October–November 1973): 22–24.

Rhodes, Randy. "Processing and Marketing Catfish." *Alabama Agribusiness* 30 (spring 1992): 2–4.

Robinson, Clay M. "The Humble Catfish." *American Fishes and U.S. Trout News* 12 (May–June 1968).

"Rotarians Hear about New STRAL Processing Plant." *Greensboro Watchman* (22 August 1968): 8.

Schmittou, H. R. "The Culture of Channel Catfish, *Ictalurus punctatus* (*Rafinesque*) in Cages Suspended in Ponds." *Proceedings of Twenty-Third Annual Conference of the Southeastern Association of Game and Fish Commissioners* 23 (October 1969): 226–44.

Schweid, Richard. *Catfish and the Delta: Confederate Fish Farming in the Mississippi Delta*. Berkeley, Calif.: Ten Speed Press, 1992.

"A Scientist's Viewpoint of the Catfish Industry." *Catfish Farmer* 5 (November 1973): 24–26.

Shell, Wayne. *The Development of Aquaculture: An Ecosystems Perspective*. Alabama Agricultural Experiment Station. Auburn University, Ala., 1993.

———. "The Role of Aquaculture in Food Production." *Alabama Agribusiness* 22 (July 1983): 1.

Shira, Austin F. "Notes on the Rearing, Growth, and Food of the Channel Catfish, *Ictalurus punctatus*." *Transactions of the American Fisheries Society* 46 (March 1917): 77–88.

Sibley, David Allen. *The Sibley Guide to Bird Life and Behavior*. New York: Alfred A. Knopf, 2001.

Smith, E. V. "Philosophical and Administrative Matters." Speech Given on the Occasion of the Signing of a Grant for the International Center of Aquaculture, June 26, 1970. Special Collections and Archives, Auburn University, Auburn, Ala.

Southern Pride Catfish Company. "Southern Pride Catfish Acquired by American

Seafoods." 21 November 2002. Select: "News." http://www.southernpride.net/ (accessed May 2005).

———. "Southern Pride Catfish Prepares for Growth and Expansion." 22 November 2002. http://www.southernpride.net/southernpride/news/American%20Seafood%20acquisition.htm (accessed 27 February 2003; page now discontinued).

Southern Regional Aquaculture Center. "Results at a Glance." Select: "Safety and Sanitation of Aquacultural Products: Microbial." http://www.msstate.edu/dept/srac/results.htm#FOOD%20SAFETY-MICROBIAL (accessed 20 November 2002).

Stepanoff, A. J. *Centrifugal and Axial Flow Pumps: Theory, Design, and Application.* 2nd ed. New York: John Wiley and Sons, 1957.

Stephens, Check (Chester O). Tape recording made at the request of Wayne Shell in the mid-1990s. Special Collections and Archives, Auburn University, Auburn, Ala.

Stickney, Robert R. *Aquaculture in the United States: A Historical Survey.* New York: John Wiley and Sons, 1996.

Sumlar, Anisa. "Farm-Raised Catfish: A Bright Spot for Alabama Agriculture." *Neighbors* 22 (September 1997): 16–17.

Swingle, Homer. "Experiments on Growing Fingerling Channel Catfish to Marketable Size in Ponds." *Proceedings of the Annual Conference of the Southeastern Association of Game and Fish Commissioners* 12 (1958): 63–72.

———. "Preliminary Results on the Commercial Production of Channel Catfish in Ponds." *Proceedings of the Annual Conference (1956) of the Southeastern Association of Game and Fish Commissioners* 10 (1957): 160–62.

Swingle, Homer, and E. V. Smith. "Management of Farm Fish Ponds." Alabama Polytechnic Institute [Auburn University]: Auburn, Ala., 1938. Mimeograph.

"Terramycin Cleared for Fisheries." *American Fishes and U.S. Trout News* 14 (November–December 1970): 27.

Thompson, E. "Alabama's Catfish Industry Here to Stay." *Neighbors* 6 (October 1981): 14–15.

Tiddens, Art. *Aquaculture in America: The Role of Science, Government, and the Entrepreneur.* Boulder, Colo.: Westview Press, 1990.

True, Richard. "A Report on the ACMA." *Catfish Farmer* 4 (March 1972): 40.

U.S. Environmental Protection Agency. *Development Document for Proposed Effluent Limitations Guidelines and Standards for the Concentrated Aquatic Animal Production Industry Point Source Category.* Chapter 4, "Industry Profiles," [sec.] 4.1.2., "Development of Commercial Aquatic Animal Production." http://epa.gov/guide/aquaculture/tdd/ch4.pdf (accessed April 2005).

U.S. House of Representatives Committee on Agriculture. "Agriculture Glossary." "National Aquaculture Act of 1980." http://agriculture.house.gov/info/glossary/n.htm (accessed May 2005).

Wellborn, Thomas L. "History of the Channel Catfish Industry." 12 February 1990. Special Collections and Archives, Auburn University, Auburn, Ala.

"Where Do We Go from Here?" *Commercial Fish Farmer* 1 (May–June 1975).

Interviews by Author. Transcripts Available in Special Collections and Archives, Auburn University, Auburn, Ala.

Allison, Ray. Opelika, Ala., 8 May 1998.
Baker, William S. Greensboro, Ala., 7 May 1998.
Boyd, Claude E. Auburn University, Ala., 14 April 1998.
Brooks, Teresa. Lincoln, Ala., 11 June 1998.
Burdette, Ron R. Oxford, Ala., 6 April 1998.
Butler, Elbert L. Childersburg, Ala., 11 June 1998.
Butterfield, Dan. Tuscaloosa, Ala., 14 January 1998.
Carlisle, Jimmy. Montgomery, Ala., 13 January 1998.
Coleman, B. Wynne, III. Greensboro, Ala., 13 May 1998.
Crews, Jerry. Auburn University, Ala., 19 February 1998.
Dunham, Rex A. Auburn University, Ala., 11 April 2000.
Easterling, William B. Clio, Ala., 2 January 1998.
Ezell, Mary H. Lavaca, Ala., 13 May 1998.
Farquhar, Andrew S., Grover Nevin, James Bonds, and Bill Mason. Greensboro, Ala., 31 March 1998.
Friday, J. V., and Ellen K. Friday. Selma, Ala., 26 February 1998.
Glover, Joe, Jr. Greensboro, Ala., 7 May 1998.
Glover, Joe, Sr. Greensboro, Ala., 27 February 1997.
Hollingsworth, Randall R. Greensboro, Ala., 26 February 1998.
Jensen, John W. Auburn University, Ala., 10 December 1997.
Kelly, Hugh D. Florence, Ala., 16 January 1998.
Kyser, Joe, Jr. Uniontown, Ala., 27 January 1998.
Kyser, William T., and Beverly Kyser. Greensboro, Ala., 19 January 1998.
Lavender, A. L. (Buddy). Boligee, Ala., 29 March 1998.
Livingston, Clark. Greensboro, Ala., 12 May 1998.
Lovell, Richard T. (Tom). Auburn University, Ala., 15 April 1998.
McCrory, Randall A. Uniontown, Ala., 27 January 1998.
Miller, Harry, and Neal Miller. Safford, Ala., 26 February 1998.
Montz, Lenson, and Ann Montz. Greensboro, Ala., 20 January 1998.
Pearce, David. Browns, Ala., 28 January 1998.
Phillips, W. S. (Shep). Equality, Ala., 12 March 1998.
Plumb, John A. Auburn University, Ala., 23 March 2000.
Prather, Ellis E. Auburn University, Ala., 5 March 1998.
Rhodes, Thomas R. (Randy). Greensboro, Ala., 1 April 1998.
Rogers, Wilmer A. (Bill). Auburn University, Ala., 15 April 1998.
Schmittou, Homer N. (Rudy). Auburn University, Ala., 13 April 1998.
Shell, E. Wayne. Auburn University, Ala., 16 April 1998.
Smelley, George. Uniontown, Ala., 27 January 1998.
Snow, Jack R. Marion, Ala., 2 April 1998.

Spree, Thed. Boligee, Ala., 21 January 1998.
Stephens, Chester O. (Check). Selma, Ala., 28 February 1997 and 18 March 1998.
Taylor, Milton, and Lester H. Taylor. Town Creek, Ala., 15 January 1998.
Thomas, Charles W. Greensboro, Ala., 30 March 1998.
True, Richard. Newbern, Ala., 26 February 1997.
Weissinger, William. Browns, Ala., 28 January 1998.
Wright, Elvin V. Greensboro, Ala., 12 May 1998.

Index

Page numbers in italics refer to illustrations.

www.ingramcontent.com/pod-product-compliance
Lightning Source LLC
LaVergne TN
LVHW050600110926
847619LV00018B/113
9780817353445